COMPUTERS IN
GEOGRAPHY

DAVID J. MAGUIRE

Computers in Geography

David J. Maguire

COMPUTERS IN GEOGRAPHY

Longman
Scientific &
Technical

Copublished in the United States with
John Wiley & Sons, Inc., New York

Longman Scientific & Technical
Longman Group UK Limited,
Longman House, Burnt Mill, Harlow,
Essex CM20 2JE, England
and Associated Companies throughout the world.

Copublished in the United States with
John Wiley & Sons, Inc., 605 Third Avenue, New York, NY 10158

© Longman Group UK Limited 1989

First published 1989

British Library Cataloguing in Publication Data

Maguire, David J.
   Computers in geography.
   1. Geography. Applications of microcomputer
   systems
   I. Title
   910′.28′5416
   ISBN 0-582-30171-8

Library of Congress Cataloging-in-Publication Data

Maguire, D. J. (David J.)
   Computers in geography / David J. Maguire.
      p.   cm.
   Bibliography: p.
   Includes index.
   ISBN 0–470–21194–6 (Wiley, USA only).
   1. Geography – Data processing.   I. Title.
G70.2.M35 1989   88–18732
910′.285 – dc 19   CIP

Set in Linotron 202 10/11pt Ehrhardt Roman

Produced by Longman Group (FE) Limited
Printed in Hong Kong

Contents

List of figures

List of plates

Preface

This book is concerned with the applications of computers in geography. It is not a computer manual; indeed it is not a computer book as such. It is a geography book which considers how geographers from all branches of the discipline can enhance their work by using computers. Little knowledge of computers and quantitative techniques is assumed at the outset. The necessary key concepts and techniques are introduced at the appropriate place within the text.

This book follows a logical progression. It begins with an introductory chapter that overviews the subject of computers in geography. The following chapters consider how computers can be used in each of the major stages on the road to geographical explanation. Thus there are chapters on data collection, storage, management, analysis and presentation. Each chapter has suggestions for further reading and there is a comprehensive bibliography at the end of the book. Technical phrases used in the text are explained in a glossary also at the end of the book. Examples are used throughout to demonstrate how best to apply information technology to specific problems in both human and physical geography. The examples are illustrated by over 100 photographs, line drawings and tables.

The general discussion is relevant to mainframe computers, mini-computers and microcomputers alike. Special emphasis is, however, given to microcomputers, reflecting the enormous contribution they have made to geographical computing in the 1980s. Many of the examples are based around Acorn and IBM hardware and software products which are in widespread use in educational establishments in several countries, notably in Britain and North America. Whilst these examples have been designed to illustrate general principles, inevitably it has been necessary to present some specific hardware and software details. It should be borne in mind when reading the examples that many other combinations of hardware and software can be used to achieve similar results.

This book is aimed primarily at first- and second-year undergraduates, though much of the material will be relevant to other students and lecturers. Undergraduates will find that it offers an overview of current and potential future applications of computers in geography. It also presents details of the impact of information technology on science and society. Lecturers should be able to use it to find ways of using computers to enhance their teaching, research and administration. Those lecturers concerned with departmental resource allocation should, in addition, find the reviews of hardware and software and the discussion about the role of computers of value.

Two frequent criticisms of books that discuss computers are that they are often out of date before they are published and that they contain too much crystal-ball gazing. To try to avoid the first criticism I have liaised closely with the publishers to minimize delays. To try to avoid the second I have restricted discussion of the latest or shortly-to-arrive hardware and software, which some cynical computer users refer to as 'vapourware', because they often never arrive or live up to their advanced publicity. I have aimed slightly on the conservative side and have chosen to discuss only those aspects of computing in geography that have been proven at the time of writing.

Last, but by no means least, I would like to thank and dedicate this book to Heather, Amy and Walt, without whom this book would not have been possible or worthwhile. One of them read, commented on and corrected everything; the others slept through most of the process.

David J. Maguire
Whitecroft, Oadby
February 1988

Acknowledgements

We are grateful to the following for permission to reproduce copyright material:

American Society of Photogrammetry and Remote Sensing and the author for fig. 6.8 from plate 3a, pp. 87–100 of Vol. 52 (1) (Jensen *et al.* 1986); Association of American Geographers and the authors for fig. 1.3 from table 1 (Bowlby & Silk 1982); Basil Blackwell Ltd. and the author for fig. 1.1 from fig. 1 (Rhind 1988); The British Petroleum Company Plc. for Plate 3 and figs. 7.10, 7.11 & 7.12 from the Slick! computer pack; the author, J. Dangermond for figs. 10.3 & 10.6 (Dangermond 1983); W. H. Freeman and Company for fig. 6.1 (Sabins 1987); the Controller of Her Majesty's Stationery Office for fig. 10.8 (DoE 1987); the Institute of British Geographers and the authors for figs. 1.2 (Palmer 1986), 1.4 (Anderson 1982), 2.2 (Sumner & Sparks 1984) & 5.5 (Vincent & Haworth 1984); Journal of Geography in Higher Education for fig. 7.2 from fig. 1 (Haines-Young 1983); Longman Group U.K. Ltd. and the author for figs. 6.2 & 6.3 from figs. 2.1 and 2.11 (Curran 1985); Longman Group U.K. Ltd. on behalf of Oliver & Boyd Ltd. for fig. 7.5 from fig. 1 (Burcham & Ferguson 1985); the author, M. A. Murray for fig. 5.4 from Map 8 (Murray 1974); National Remote Sensing Centre for Plate 2 and figs. 6.6a, 6.6b and 6.6c; Nigel Press Associates for figs. 6.6d & 6.7; Pergamon Journals Ltd. and the authors for figs. 2.5 & 2.6 (Clarke, Fisher & Ragg 1986), 5.8, 5.9 & 5.13 (Jones 1985); Pion Ltd. and the authors for figs. 7.8 & 7.9 from figs. 2 & 3 (Batty & Longley 1986); Pitman Publishing Ltd. for figs. 11.2 & 11.4 from figs. 2.1 & 2.8 (Shelly & Hunt 1984); Taylor & Francis Ltd. and the author, R. J. Whittaker for figs. 10.12 & 10.13 from figs. 2 & 4 (Wiggins, Hartley, Higgins & Whittaker 1987). Also Mr B. W. Hickin and Dr A. J. Strachan for Plate 1.

Whilst every effort has been made to trace the owners of copyright material, in a few cases this has proved impossible and we take this opportunity to offer our apologies to any copyright holders whose rights we may have unwittingly infringed.

1

Computers in Geography

Computers may be defined here simply as electronic machines capable of the input, storage, manipulation and output of data (a much fuller description and definition is presented in Chapter 11). Computers are a very important part of science and society. They form the basis of control systems for a diverse range of operations which includes aircraft flight, electricity flow and security systems. They are used to collect information about the environment, store information about people, manipulate statistical data and draw maps and graphs. Today very powerful supercomputers can work at speeds of 1,000 million instructions per second and can store 32 million words in main memory. Linked to computers, graph plotters can draw lines at 1 metre per second and line printers can print 500 characters per second. Small portable microcomputers, with power greater than that of the computers used to control the first space flights, are now within the price range of most people in the western world.

In spite of these impressive statistics, it needs to be stated at the outset that there is really nothing magical about computers, although to the uninitiated this may not seem so. Computers are simply tools; in a sense they are like chart recorders, peat borers, compasses and slide projectors. They have become so widely used because of the great improvements in efficiency and effectiveness that they bring. In theory, there is nothing that a geographer can do with a computer that cannot be achieved by other means. In practice, however, as will be demonstrated in this book, there are many operations which would not be contemplated without the aid of a computer, because they are too time-consuming. It is the ability of computers to perform simple repetitive operations very quickly and to store vast quantities of data that has made them so useful. Any application which can make use of these facilities can invariably be greatly enhanced when computerized.

This book is concerned principally with the applications of computers in geography. It is not a computer manual, indeed it is not a computer book as such. It is a geography book which considers how geographers can use computers to enhance their activities. Examples have been carefully selected to demonstrate how best to apply information technology to a range of specific problems in both human and physical geography.

The remaining part of this first chapter provides a general introduction to geographical computing. There are then nine chapters which discuss the major applications of computers in geography. These are concerned with data collection, data management, statistical analysis, computer cartography, remote sensing and image analysis, simulation, word processing, communication and geographical information systems. They show how computers can be used to assist geographers at all stages in their attempts to describe, explain and predict the spatial patterns and processes on the surface of the earth. Many of the exciting new developments in geographical computing are discussed at several points in the text, most notably in Chapter 10 which is concerned with geographical information systems. Following this, there are two chapters that discuss the key elements and functions of computer hardware and software. They are designed for those interested in learning more about the potential limitations and suitability of computers for geographical work and for those interested in developing new applications. A final chapter brings together some of the major themes explored and offers some details about likely future developments.

The examples presented are wide ranging and deal, amongst other things, with collecting weather station data using a data logger, creating a data base of census data, statistically analysing rainfall data, mapping social survey statistics, processing remotely sensed pollution data, simulating the structure of urban form, preparing a word processed report, accessing on-line data bases and using a geographical information system for health care planning. After reading this book students should have a greater appreciation of the current and potential future uses of computers in geography. They should be familiar with the basic details of the major applications of computers in geography and will, therefore, be better placed to understand and evaluate geographical literature which incorporates computer work. Finally, students should have a general grasp of the key issues involved in the information technology revolution which has already dramatically affected our lives.

THE QUANTITATIVE AND COMPUTER REVOLUTIONS IN GEOGRAPHY

Although computers were available commercially from the early 1950s, it was almost a decade before they made much impact upon geography. During the late 1950s and early 1960s there occurred in geography important philosophical and methodological changes which have subsequently been termed 'the quantitative revolution' (Johnston 1987). The revolution heralded a change in geography, from an essentially qualitative and descriptive discipline into one which became increasingly concerned with the development of generalized laws and theories about spatial patterns using mathematical and statistical methods. The computers of the late 1950s and early 1960s were expensive, scarce and were difficult to use, nevertheless they still attracted the interest of geographers. American geographers were the first to use computers, Garrison (1959) and Tobler (1959) used them in their work on rural poverty and cartography. In Britain computers were used by Moser and Scott (1961) in a study of British towns and by Coppock (1964) in an agricultural atlas of England and Wales (Dawson and Unwin 1976).

It was not until the middle 1960s when more advanced computers became available at a lower price that they had much impact on mainstream geography. After a slow start the uptake of the new technology was very rapid. Computers were first used extensively for statistical analyses, such as cluster analysis, factor analysis and regression. Later, computer cartography, simulation and remote sensing all became important applications. Whether the developments in computers caused the quantitative and theoretical revolution or merely enabled it, is for others to argue. It is clear, however, that computers played a significant part in these important changes.

Twenty years on from the quantitative and computer revolution, geography once more underwent important changes. The introduction of relatively low-cost microcomputers into society in the late 1970s and 1980s and their pervasion into geography initiated a second computer revolution. Two of the earliest papers in the geographical literature which used microcomputers were published in 1979 by Chalmers, Thompson and Keown and by Degani, Lewis and Downing. Both sought to demonstrate the value of these low-cost, interactive machines for simulation work. Chalmers, Thompson and Keown used a computer to simulate an agrarian economy and Degani, Lewis and Downing used one to simulate the process of soil erosion. The microcomputer revolution stimulated renewed interest in many geographical computer applications and encouraged the introduction of new types of data loggers, word processing and communication.

3

Up until the late 1970s, the major computer problem for geographers was a lack of hardware (Fig. 1.1). Only with the increased access to computers, brought about by the installation of centralized mainframe computers and minicomputers in institutions during the 1970s and the microcomputer revolution of the 1980s, was this problem alleviated (Unwin 1974; Dawson and Unwin 1984). With the hardware problem taken care of, the lack of 'geographical' software was then identified as the major computing problem. The dearth of good quality commercial software and lack of the necessary organization for disseminating software developed by academics, led many geographers to become disillusioned with the new technology. It is only in the last few years that a reasonable amount of good-quality commercial software has started to become more widely available. This, combined with a steady flow of software developed by geographers, has served to ameliorate the software problem and its significance should diminish still further in the next few years.

There appear to be two major computing problems facing geo-

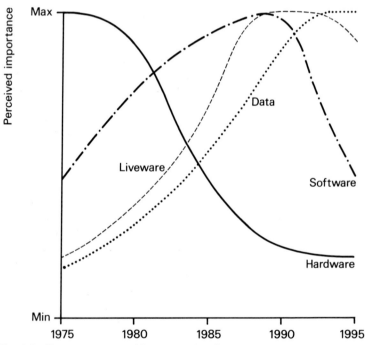

Fig. 1.1. Key computing problems in geography.
Source: adapted from Rhind 1988.

graphers in the later 1980s and early 1990s, namely, the lack of 'liveware' and the proliferation of data. The lack of liveware, that is suitably qualified people to maintain computer systems and to develop and teach the geographical applications of computers, has been identified for sometime (Dawson and Unwin 1984). It is only in the past few years, however, that it has risen to prominence. This problem can easily be overcome if sufficient funds are made available to train technicians, research students and teachers and lecturers. The proliferation of data is the latest of the computer problems to confront geographers and is more significant. It is one consequence of the information revolution which accompanied the technological revolution of the 1980s. The solutions to the problem remain to be discovered, but they must include improving data storage and processing hardware as well as the techniques of data storage

THE APPLICATIONS OF COMPUTERS IN GEOGRAPHY

The applications of computers in geography are legion. Contrary to popular belief, there are few areas of the discipline which have not already been substantially affected by the computer revolutions. It is impossible to summarize the applications of computers in geography adequately and so the aim here is to offer some examples, which will whet the reader's appetite and will provide an introduction to the themes developed in subsequent chapters. For convenience, three examples of the applications of computers in geographical research will be discussed first and then the subject of Computer Assisted Learning (CAL) will be introduced.

Computerizing Domesday Book

Historical geography, because of the incompleteness and variability of the data and nature of the research methodology, is often assumed to be beyond the influence of computers. This is in fact incorrect, since computers may be used in many aspects of historical geography. The recent advances in geographical information systems (see Chapter 10) for example, which will allow the integration of data from different time periods that are recorded using different spatial bases, promise to have a profound impact in the future. One example of a historical geography project that has been using computers for some time concerns computerizing the Domesday Book (Palmer 1986). Domesday Book is a unique survey that celebrated its 900th anniversary in 1986. It contains in excess of one million words and combines a register of land usage, a tax register and a census of people and animals. The significance of Domesday Book, the size of the volume, the complexity and structure of the record and variety of phraseology used by its scribes, have made it a prime candidate for computerization.

There are basically two strategies for computerizing a text such as Domesday Book. First, all the statistics may be manually extracted from the text and then coded into a data matrix suitable for input into one of the standard statistical analysis packages, such as SPSS or Minitab (see Chapter 4). The major advantages of this approach are its simplicity and speed. However, it has the disadvantages that much of the text is disposed of in the process and the coding is heavily dependent upon the interpretation of the coder. Alternatively, the whole text of the book may be computerized in total. The main disadvantages of this approach are that it is very time-consuming and that it requires vast amounts of computer storage. Several attempts at the former have already been completed and the latter is currently in progress.

Once the massive task of computerizing is complete, the data can be processed in a number of ways. For example, it is possible to search the date base for every occurrence of, say, a place or a person, to compare and classify areas statistically by deriving indices of wealth, land charac-teristics and population totals, to map areas and so examine spatial variations in phenomena such as crop types, land ownership and changes in land value (Fig. 1.2) and even to compare the internal logic of the book itself.

Statistical analysis of shopping survey data by computer

Geographers frequently wish to analyse data sets in which the nature of many of the variables is qualitative (categorical), i.e. counts in a number of categories, rather than quantitative (continuous). Examples of categorical variables include vegetation cover recorded as presence/absence, glacial erratic rock type recorded as granite/limestone/schist/greywacke and household tenure recorded as owner-occupied/council rented/housing association. The statistical procedures necessary to examine large sets of categorical data are, by the standards of many geographers, complex and time-consuming to implement. The de-velopment of the GLIM (Generalized Linear Interactive Modelling) computer package has done much to increase the awareness of the potential usefulness of categorical data amongst geographers (Wrigley 1985). It has also provided a relatively sophisticated, easy to use, quick and reliable method of undertaking analyses.

Bowlby and Silk (1982) present an introduction to the use of GLIM for categorical data analysis in geography, using two examples based on shopping survey data, the first of which will be outlined here. The data

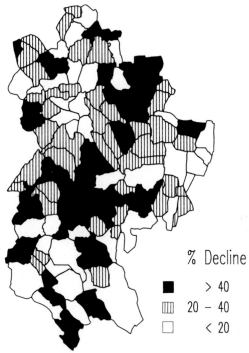

Fig. 1.2. Domesday values in Bedfordshire, showing the percentage decline in values from 1066 to 1070, derived from a computerized version of Domesday Book.

Source: Palmer 1986.

were derived from a survey of 680 people whose response to the statement 'I find getting to grocery shops very tiring' was coded on a five-point scale: 1: disagree, 2: tend to disagree, 3: in-between, 4: tend to agree, 5: agree. Details were also recorded about the respondents life-cycle stage and car availability (Fig. 1.3). Bowlby and Silk were interested in determining the effect of life-cycle and car availability on whether people find getting to shops tiring. Because both the dependent variable (level of agreement) and the two explanatory variables (life-cycle stage and car availability) are categorical, traditional techniques of regression and analysis of variance cannot be used here. Instead, a closely related statistical modelling procedure, known as log-linear modelling, must be used. The aim still remains, however, to predict statistical variation in one dependent variable using two explanatory variables. This analysis can be undertaken using just eleven GLIM commands and from start to finish takes less than 15 minutes. Accord-

7

Agreement with the Statement 'I find getting to grocery shops very tiring,' by Car Availability and Life-Cycle stage

	Car Availability	Agreement					Row Totals
		1	2	3	4	5	
Middle-	1	22	1	9	6	36	74
Aged	2	29	3	7	10	41	90
	3	27	3	6	1	24	61
Younger	1	14	8	4	6	22	54
People	2	26	2	5	2	13	48
without Children	3	27	7	3	3	16	56
Younger	1	19	2	3	5	42	71
People	2	46	2	6	11	49	114
with Children	3	37	10	16	12	37	112

Fig. 1.3. Categorical shopping survey data statistically modelled by computer. *Source*: Bowlby and Silk 1982.

ing to Bowlby and Silk, the results show that lack of a car seems to have most impact on people with children or those whose physical health may be affected by age. Young people without children tend not to find the journey to shops tiring.

Computer modelling of hillslope hydrology

Anderson (1982) describes the use of a computer model to simulate soil water conditions during drainage, on a range of hillslopes having different soil types and topographies. Previous empirical work has determined that the key factors affecting soil water conditions on hillslopes are the hydraulic conductivity of the soils and the slope angle. Instrumented hillslope sites can provide an important basis from which to begin estimating the threshold values of these parameters responsible for changing soil water conditions. However, the large number of parameter requirements has led Anderson to a less time-consuming and less expensive approach based on computer simulation modelling.

The computer simulation model utilizes a grid cell structure. It comprises a collection of 63 (9 × 7) 10-m square cells in plan view and deals with the upper 1 m of the soil. These cells form the compartments within which soil water conditions are simulated. Before the model was actually used to assess variations in hydraulic conductivity and slope angle, it was empirically verified by comparing the predicted variations in soil moisture conditions with actual data obtained from an instrumented catchment near Bristol, UK. Only at very high values of soil moisture were there discrepancies outside acceptable limits and this occurred in less than 10 per cent of the area covered.

Over 100 computer simulations were undertaken in total, with the slope angle varied from 2° to 20° and saturated hydraulic conductivity varied from 10^{-3} to 10^{-1} cm s^{-1}, over a 1,000 hour (42 day) time period. Figure 1.4 shows computer-drawn graphical representations summarizing some of the results. These results would have taken hours to calculate by hand and are manually very difficult to represent in a three-dimensional form. They demonstrate the importance of the unsaturated zone in controlling soil moisture discharge from hillslopes. Only in cases where both the hydraulic conductivity and slope angle were high does soil water always flow into hollows. For all other conditons simulated, more complex patterns of cross hollow-spur movements were shown to occur.

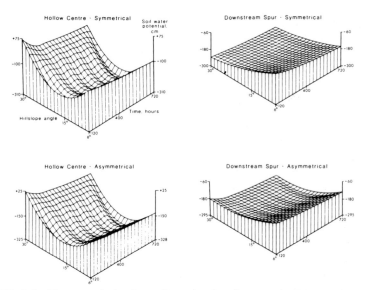

Fig. 1.4. Computer drawn three-dimensional surfaces summarizing output from a computer simulation of soil water potential on hillslopes. The saturated hydraulic conductivity is set at 10^{-3} cm s^{-1} in all cases.
Source: Anderson 1982.

Computer Assisted Learning (CAL)

The field of *Computer Assisted Learning (CAL)* is infested with jargon, such as Computer Aided Instruction (CAI), Computer Assisted Training (CAT) and Computer Managed Learning (CML), which only serves to mystify the subject further (Shepherd, Cooper and Walker 1980). The term CAL is frequently used as an umbrella to describe the full range of applications of computers as a support to teaching and learning activities. In geography, however, it is usually restricted to teaching with computers rather than about computers and it is in this sense that it will be used here. CAL in geography is not new; it has been around since the 1960s, although the expansion in computing in the early 1980s (as a result of the introduction of microcomputers) produced a great upsurge in interest in its usage (see for example Adams and Jones 1983; Kent 1983, 1987; Watson 1984a, b). Shepherd (1985), in his excellent review of CAL, considers the educational roles of CAL in relation to the teaching methods commonly used by geographers and this approach will be adopted here.

Computers can enhance both the presentation and preparation of lectures. During lectures they can be used to run simulations (see Chapter 7), to animate static maps or diagrams and to recall, analyse and display data (see Chapters 4, 5 and 6). In lecture preparations they are useful for, amongst other things, producing handouts and reading lists (see Chapter 8), helping to create study materials (by, for example, extracting data from a data base) and providing indexing and retrieval facilities for teaching resources. Computers may be able to assist and encourage students to read more and, more importantly, to identify relevant material. A number of bibliographic search and retrieval systems have already been developed for such purposes and several of these are discussed in Chapters 3 and 9. The recent developments in Desk Top Publishing (see Chapter 8), Videotex (see Chapter 9) and electronic books, which have facilities for browsing and annotation, also offer potential for structured reading.

Lecturing and reading are both 'passive' mass consumption teaching methods which offer limited scope for CAL. In contrast, tutorials and fieldwork are both 'active' methods which can easily respond to the needs or wishes of the individual learner. Small group teaching of this nature lends itself to CAL methods such as simulation games, data analysis and display, problem solving and hypothesis testing. Most microcomputers are sufficiently portable to enable them to be taken to residential field centres (Gardiner and Unwin 1986) and some can be used for field data logging (see Chapter 2). Using computers the statistical analysis and display of data can be undertaken on the field course whilst student minds are still attuned to the problem under study. The advent of videodisk systems, like the Domesday system (see Chapter 10), may offer students the opportunity to prepare themselves more fully. Such videodisks may even one day replace the 'Cook's Tour' so common on the first day of a week-long field course.

Finally, CAL can enhance students' written work. Essays, laboratory and field reports are still the basic method of continuous course assessment in most geography departments. A word processing package (Chapter 8) can greatly assist students in preparing and redrafting manuscripts. Given practice with a word processor, student writing skills can be greatly enhanced because of the added opportunities to experiment with layout and style.

It is tempting, particularly for those involved in its everyday use, to oversell CAL. It is probably best viewed simply as one more teaching method to be added to a teacher's armoury, which can be used to teach certain things, to certain people, at certain times. Little is actually known about the precise benefits of CAL in geography, since there is little objective systematic analysis of its relative merits. Some educationalists believe that using CAL will greatly increase the efficiency of teachers. However, the available data suggest that the creation of CAL software can be enormously expensive, both in terms of time and money. This is in addition to time spent learning how to use a particular computer system. The amount of CAL in current use is still sparse and is far below the critical mass needed to make a major impact on learning. Many of the materials which are available are little more advanced than some of those developed in the 1970s.

DEVELOPING COMPUTER APPLICATIONS IN GEOGRAPHY

There is currently an enormous variety of computer hardware, software and data available to geographers. The choice of which to purchase or use is a difficult one that is rather dependent on local circumstances, although there are a number of generally applicable factors that should be considered. The first thing to decide is what the new hardware, software or data are required for. The more closely the requirements can be defined, usually the easier it will be to fulfil them. Next a short list of possible contenders should be established by talking to current users, reading reviews (many contain benchmarks which can be a useful guide) and manufacturers' specifications. Demonstrations should then be arranged which, preferably, directly demonstrate or mimic the actual application for which the goods are required. The amount of after-sales service and the cost of maintenance contracts and training (if available) should be ascertained. It is a good idea to examine the quality of the documentation if similar items have not been used before. Generally speaking, it is better to choose items which have been on the market for some time that have been extensively tried and tested. Lastly, with so many imitations of popular hardware and software available it is often difficult to decide whether to buy the original or a so-called higher performance and/or lower price compatible.

Because of the time and financial costs involved in developing software it makes sense, where possible, to purchase good quality proprietary software. Unfortunately, it may be difficult to find a piece of software which exactly fits an application and in such cases it may be necessary to write bespoke software. A number of authors offer guidelines for designing and developing software (see for example Burkhardt *et al* 1982; Chan and Korostoff 1984; Midgley and Walter 1985; Unwin and Dawson 1985). Unwin and Dawson (1985) suggest that well written software should have a number of characteristics. It should be well designed and written in a portable language, so that it can run on more than one machine type. The code should be well presented and clearly annotated, so that it is easy to follow. The software should be user-friendly and should be accompanied by good documentation, so that it is easy to use. Lastly, the software should give the correct answer (behave in the manner intended by the author) and should be programmed efficiently to minimize run time and processor time (see Chapters 11 and 12).

Writing computer software to solve a geographical problem involves a number of important steps. The first step is to define the problem. This generally necessitates breaking it down into small units. It is worth considering at this stage whether anyone else has written software which solves part or all of the problem and if so whether it can be used. Once the problem has been clearly defined it can be converted into an algorithm, which is a clearly defined solution to a specific problem. Some people like to use flow charts, that have special symbols, to outline the course of machine control through a sequence of steps. Others prefer to sketch out the problem in pseudo-code (abbreviated code showing only the main steps). When a suitable algorithm has been devised it is necessary to begin coding. This relatively small and easy part of the operation is what most people believe to be computer programming. Once a problem has been completely coded, it can be entered into a computer. The software program must then be exhaustively tested and debugged to ensure that it is user-friendly, computationally correct and efficient. Last, appropriate documentation must be written to accompany the software. This should include details of the problem, the algorithm, a program listing and a specimen data set.

The characteristics of data are an all-too-often neglected area of computer-based geographical investigations. The most important points to note are data availability (Rhind 1985), quality (Chrisman 1983; Blakemore 1985) and quantity (Rhind 1983). Government agencies play an important role in collecting and disseminating data for much research in human and physical geography (DoE 1987). Well used examples include the agricultural and population censuses, meteorological and pollution data. Some data, such as environmental remote sensing satellite data and topographic digital map data, can be extremely expensive. The cost of these secondary sources is, however, substantially below that of primary data collection in the majority of c̲ es. The restricted availability of data in Britain contrasts markedly with several other countries, including the United States, where many large data sets are in the public domain and only small charges are made for copying them.

The lack of information about the quality of digitized outlines and geographical data bases has been identified as a key problem in recent years. Such information is necessary so that users can make an informed judgement about the suitability of data for a particular purpose. Although all conventional mapping agencies have published map accuracy standards, these have not been developed for computer cartography. Chrisman (1983) suggests that information about the quality of feature recording, digitizer reliability, positional accuracy and the age of a map are all important aspects of data quality. Standards for data quality are also important prerequisites for a more united approach to data integration in geographical information systems.

The very large size of many spatial data bases, such as remote sensing satellite images and digital map data files, is a further problem which computer users need to overcome. These create problems both for storage and processing. The advent of new hardware devices will continue to assist in the process, but at present the rate of data collection far outstrips the technology. New efficient data storage and processing techniques are, therefore, a matter of some urgency.

CONCLUSION

This chapter has overviewed the subject of computers in geography. The characteristics and capabilities of computers have been briefly described and their introduction into geography has been discussed. The four examples considered were chosen to illustrate the great wealth of applications of computers in geography. These examples also serve to introduce several of the themes developed in the next nine chapters which deal with the major applications of computers in geography. These chapters have been arranged in a logical progression that shows how computers can be used in each of the major stages on the road to geographical explanation. The next chapter, therefore, is concerned with the collection of geographical data.

FURTHER READING

Johnston, R. J. (1987) *Geography and Geographers: Anglo-American Human Geography since 1945* 3rd edn. Edward Arnold. (Discusses the background to the quantitative revolution in geography.)

Midgley, H., Walker, D. R. F. (1985) *Microcomputers in Geography Teaching*. Hutchinson. (Some useful remarks about developing software on microcomputers, but little on background details and wider issues.)

Shepherd, I. D. H. (1985) Teaching geography with the computer: possibilities and problems. *Journal of Geography in Higher Education* **9**:3–23. (An excellent introduction to CAL in geography which nicely summarizes the key issues and practical applications.)

Shepherd, I. D. H., Cooper, Z. A., Walker, D. R. F. (1980) *Computer-assisted Learning in Geography: Current Trends and Future Prospects*. Council for Educational Technology with the Geographical Association. (A very useful book on CAL in geography. Much of the discussion is still pertinent today.)

Unwin, D. J., Dawson, J. A. (1985) *Computer Programming for Geographers*. Longman. (A more discursive book than the title suggests that provides the background to computers in geography.)

Geographical data collection

This chapter is concerned with the collection of geographical data, one of the most important phases of any geographical investigation. The amount and type of data collected and the way they are stored may condition the type of analysis that can be performed and, therefore, the results obtained. Data collection is often the most time-consuming and costly phase of a project, especially if it is undertaken in remote and inaccessible locations. It is also important to pay attention to data quality during the data collection phase of an investigation, because errors introduced at this stage will remain throughout and may be compounded by further analysis.

In this book, following the recommendation of Date (1981), clear distinction is made between the terms *data* and *information*. The term data refers to the values physically recorded by an observer and stored in a computer and information refers to the meaning of those values as understood by some user.

Prior to the availability of computers, data collection and analysis involved the manual recording of data using pencil and paper and then hand calculation of the results. With the wider use of computers, it soon became obvious that data analysis could be both speeded up and made more sophisticated. Computer analysis requires data to be coded into a format suitable for input into a computer. In early investigations, data collected either by manual means or by machine were manually coded into computer format and then manually entered into a computer for analysis, usually via punched cards or a keyboard. Unfortunately, this type of data entry has proved to be time-consuming, tedious and error prone, and should be avoided if at all possible. In an attempt to try to automate the data collection process, a number of specialist data loggers have been developed which can collect data and then directly transfer them into computers. Attempts have also been made to speed up and improve the reliability of manual data entry, by using new devices and by developing software to assist with keyboard entry.

This chapter begins with a basic introduction to geographical data. The next section discusses the basic operation of data loggers and gives some examples of their application in geography. This is followed by discussion of developments in the area of manual data entry using specialist devices and keyboards. Several of the other chapters in this book also contain discussion of specific purpose data-input devices. For example, digitizers are discussed in Chapter 5 and video cameras in Chapter 6.

GEOGRAPHICAL DATA

Geographical data may be obtained from both primary and secondary sources. *Primary* sources of data include field observations such as personally conducted questionnaires, telephone interviews, field notes of stratigraphy and instrument readings. *Secondary* (also called derived) data sources include published or archival material such as population censuses and remote sensing satellite imagery. Further details of the principal sources of geographical data and discussion of the statistical aspects of data collection and sampling methods may be found in Shaw and Wheeler (1985).

The data which geographers collect may be classified in a number of different ways depending, principally, upon the purpose for which the data are to be used. Since the various types of data are stored and manipulated differently by computers, it is worth outlining the main types of data which geographers use. One of the most useful classifications is that first put forward by Stevens in 1946 (Unwin 1981). Stevens identified four basic levels of measurement that depend on the amount of information associated with data values. In ascending order the levels of measurement are: nominal, ordinal, interval and ratio. *Nominal* data have only sufficient information to classify them into categories. For example, rocks can be classified as granite, limestone, schist etc. *Ordinal* data contain sufficient information so that they can be ranked in ascending or descending order. For example, some social classifications seek to classify households by the occupation of the household head. People in professional occupations are usually placed in class 1, semiprofessional in class 2, manual in 3 etc. *Interval* data have the property that distances between categories are defined as fixed equal size units. Thermometers, for example, measure temperature on an interval scale, ensuring that the difference between, say 20 °C and 25 °C, is the same as that between 0 °C and 5 °C. However, because the scheme lacks a fixed zero only differences and not absolute values can be measured. *Ratio* data have in addition an absolute zero. A value of 0 mm of rainfall indicates no rainfall, whereas 0 °C does not indicate no temperature. It is also possible to calculate ratios from data measured at the highest level. For example, 1,000 mm is twice as much as 500 mm, but it is not sensible to say that 50 °C is twice as warm as 25 °C.

Many geographical data also have some type of spatial component associated with them. The conventional geometrical classification, based on the dimensionality of length of the data, can be suitably used here. In this scheme points have no length (l) dimension and are said to have a dimensionality of l^0. Lines have a single length dimension and are given a dimensionality value of l^1. Areas have two length dimensions and, therefore, have a dimensionality value of l^2. Finally, surfaces have three length dimensions (height being the third) and are given a dimensionality value of l^3.

For completeness, it is worth mentioning that geographers who wish to examine how spatial patterns and relationships change through time utilize temporal (also called time) data. The temporal component inherent within such data also has a strong influence on how they are collected, stored and analysed.

DATA LOGGERS

Data loggers are specialist devices that can be used to automate the process of collecting and recording data. Both automatic and semiautomatic data loggers are used in geography for a wide variety of applications. *Automatic data loggers* are those which require virtually no human intervention once they have been programmed and installed in the field. In contrast semiautomatic data loggers require at least some human intervention after installation. Although often controlled by microprocessors, data loggers are only capable of carrying out a limited number of specialist functions and because of this they may not be considered true computers. However, because data loggers have already become widely used in geography and because they are often linked to computers, it seems worthwhile outlining briefly the basic principles of their operation and describing some of their applications in geography.

A number of specialist automatic data loggers have been developed for collecting geographical data. Some of these incorporate one of the many general purpose automatic data loggers which are currently available, such as the Squirrel and Harvest data loggers. At Plymouth

Polytechnic, UK, for example, a Squirrel data logger, linked to a network of tensiometers, has been used to collect data about soil moisture status in the Plynlimon catchment in upland Wales. Further examples of the applications of specialist automatic data loggers in geography include, the collection of precipitation data (Burgess and Hanson 1983), the acquisition and analysis of field tillage data (North 1983), the recording of electrical conductivity in proglacial streams in Switzerland (Fenn 1987), and weather station monitoring (see below).

Semiautomatic data loggers are also widely used by geographers for data collection. At the University of Leicester, UK, for example, a Sedigraph linked to an Acorn BBC microcomputer is used to collect sediment particle size data. Further examples of the use of semiautomatic data loggers in geography include, the balance interface system discussed below and the digitizer for collecting locational data discussed in Chapter 5.

The use of data loggers in geography has both advantages and disadvantages compared to traditional manual methods (Armstrong and Whalley 1985) and these will be discussed below. There are four main advantages, although they do not apply equally to every type of data logger:

1 Using fully automatic data loggers, data collection can continue unattended for long periods of time, thus reducing problems of site remoteness or inaccessibility.
2 If more than one logger is available several sites can be covered at the same time.
3 Loggers can save the cost of data entry and avoid the accidental introduction of errors. Thus they can reduce the total data processing time of a project.
4 By using microprocessors to control the logging operation, it is possible to install 'intelligent' recorders capable of, for example, switching between sampling strategies and undertaking on site data correction or reduction.

There are four main disadvantages, although again they do not apply equally to every type of data logger:

1 Unfortunately, data loggers are generally costly. The relatively high cost of even simple commercial units is difficult to justify for many projects and programmable microprocessor-controlled 'intelligent' data loggers, that cost ten times the price of simple units, can only be used where they are considered essential.
2 Data loggers can generate vast amounts of data which need an appropriate amount of computer storage space and processing time.

3 Most data loggers are quite heavy and bulky, largely because of the necessary batteries and robust casing. Many cannot be carried for long distances and if left in the field they are difficult to hide. This is less of a problem than it used to be since the advent of microprocessors has led to a miniaturization of computer systems.

4 It is often difficult to see if a data logger is functioning correctly in the field. Although some data loggers have small visual display units, it is not usually possible to check if the data are being stored correctly until the storage medium is returned to the laboratory.

Data loggers may be classified most easily according to the characteristics of the data they output. Analogue data loggers, such as temperature sensors and chart recorders, produce a continuously varying signal which must be digitized before it can be read by a computer. The principles of analogue to digital conversion are discussed in Chapter 9. A number of instruments are available which emit a binary signal coded using the Gray system (named after its inventor). Some wind-vanes, for example, output data codes which correspond to different wind directions (N might be coded 000, NE 001, E 002 etc.). However most binary data, from devices such as digitizers, hand-held portable terminals and optical character readers, are coded using the ASCII system (see Chapter 12 for an explanation of this term). Event recorders, such as stream gauges and rainfall recorders (both of which use tipping buckets) and traffic recorders, emit a signal for each event.

Using an automatic data logger for weather station monitoring

The normal method of recording meteorological information is manual. At set times everyday, visits are made to a weather station, instruments read and data recorded (Meteorological Office 1982). Frequently the data are entered into a computer for storage and analysis, and various suites of programs have been written to assist with this process. Sparks and Sumner (1982), for example, describe a suite of BASIC programs for analysing such data using a microcomputer. The number of visits to a weather station can be reduced by using a preprogrammed data logger to collect and store data on a magnetic disk or tape, for input to a computer for analysis (Sparks 1983). Although an improvement on manual data collection, such systems tend to be inflexible since analysis times must be predefined, there are significant delays in processing time and it is not possible to check if data collection is proceeding correctly whilst at the weather station.

Sumner and Sparks subsequently developed a microcomputer-based system for the collection, storage, analysis and display of weather station data (Sparks and Sumner 1984a, b; Sumner and Sparks 1984). The system installed at the University of Wales, Lampeter, UK, fulfils many of the functions of more expensive commercial weather station systems, such as those marketed by Delta T and Campbell Scientific. It consists of three basic elements. Firstly, for collecting data, there are sensors including a thermometer, a wind-vane, and an anemometer (for measuring wind speed). These output, respectively, analogue, binary and event data. Secondly, for controlling data collection, storage, analysis and display there is an Acorn microcomputer with a disk drive, screen and dot matrix printer (Fig. 2.1). Thirdly, there is an interface device between the sensors and the microcomputer, called the MODAS (MOdular Data Acquisition System). This conditions the raw signals from the sensors into a form acceptable to the microcomputer and carries out some preprocessing of data.

At Lampeter the sensors are in the weather station. These are connected by cable to the MODAS and then the microcomputer, both of which are 60 m away in the laboratory. Microcomputer software, written in BASIC, is used to make the MODAS obtain data from a sensor. The MODAS processes the data and transmits them to the microcomputer where they are stored. The software also provides options for analysing and displaying data. Examples of dot matrix printer graphs are shown in Fig. 2.2.

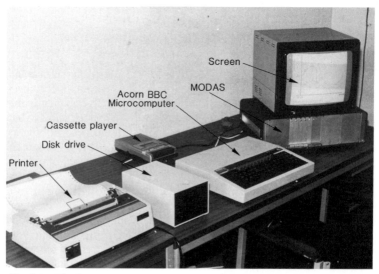

Fig. 2.1. The Lampeter weather station monitoring system.
Source: provided by L. Sparks.

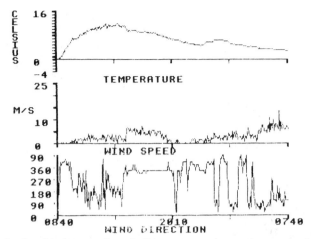

Fig. 2.2. Graphical output from the Lampeter weather station monitoring
system showing combined 5-minute plots for 9 April 1983.
Source: Sumner and Sparks 1984.

The weather station monitoring system has been used successfully at
Lampeter for several years. The data displayed in Fig. 2.2, for the 9
April 1983, clearly show that in mid-afternoon there was a fall in
temperature, accompanied by an increase in wind speed from about 3 m
s^{-1} to 5–6 m s^{-1}, and a veer of wind direction from south to north-west.
According to Sumner and Sparks (1984), these changes indicate the
arrival of a sea breeze at Lampeter. The apparent wide fluctuation of
wind direction at times of light winds is thought to be a function of the
poor local exposure, the poor low velocity response of the wind-vane
and partially the scaling of the y-axis of the graph.

Using a semiautomatic balance data logger

Many of the laboratory techniques used in geography, such as the
determination of particle size and moisture content of sediments and
soils, necessitate weighing samples. Calculations are often performed
on weights so that results can be expressed in a standard form. In the
case of soil particle size analysis, the raw weights of soil, minus the
weight of a container, frequently need to be expressed in the form of the
percentage sand, silt and clay in a sample. It may also be desirable to plot
the data as a histogram or line graph, or calculate some summary
statistics. Any data logging system which increases the speed and reliability of data collection will obviously be beneficial to the laboratory
scientist.

Fig. 2.3. The Oertling balance – Apple II microcomputer semiautomatic balance interfacing system in operation in a laboratory.
Source: provided by S. Ross.

The system described here (Fig. 2.3) allows an Apple II microcomputer to communicate with any four-line output Oertling electronic balance. Weights can be transferred from a balance to a microcomputer where software can store, analyse and display data and final calculations. The only specialist pieces of hardware required (in addition to a balance and microcomputer) are a balance interface card, plugged into the printed circuit board of the microcomputer, and a lead forming a direct communciation link between the balance and the microcomputer. The transfer of data is controlled by software which is supplied by the manufacturer, but users are required to write their own programs for data storage and analysis. The data transfer software consists of an eighteen-line BASIC program in three parts. The first part initializes the data transfer operation, that is, defines which channel the data are to be read from. The second part reads the data from the balance down the defined channel, using a special sequence of commands (called handshaking – see Chapter 9) to control the rate of transfer. The third part prints the data on the computer screen. This system can easily be used for a range of applications including particle size analysis, measuring the moisture and organic matter content of soils, and determining the sediment content of water samples.

The system is relatively simple, effective and cheap. It has the potential disadvantage of requiring a dedicated computer for the duration of the weighing operation. However, it is precisely this type of operation that can effectively utilize technologically redundant, but mechanically and electronically serviceable machines.

DATA ENTRY

The manual entry of data into computers remains important today for a number of reasons. It is still the cheapest, fastest and simplest method of data input for small one-off jobs of, for example, a few hundred observations. On some occasions it must of necessity be used, since reasonably priced automatic and semiautomatic data loggers have not yet been developed for all applications. For example, it is difficult to see how a data logger could be developed using current technology to automate the process of recording the responses to an unstructured social survey interview. The process of manual data entry into a computer may take place either indirectly, using some form of specialist data entry device, or directly using a keyboard.

Specialist data entry devices

A number of specialist devices (such as hand-held portable terminals) have been developed to assist the process of manual data entry. The Ferranti Market Research Terminal, for example, is in essence a specialist manual data entry device for the field collection of social survey data (Rowley 1985; Rowley, Barker and Callaghan 1985). The unit is designed in the form of a clipboard and holds the questions to be asked on paper (Fig. 2.4). However, with the use of a keypad, the answers to the questions are fed directly into the preprogrammed memory which is maintained by a battery. At the end of an interview or when the memory is full, the data can be transferred to a computer using a communication link (see Chapter 9). This type of system is obviously the forerunner of things to come, though at present these operations could be carried out more cheaply by some lightweight portable microcomputers.

Keyboard data entry

Direct manual data entry using a keyboard is the simplest and most common form of data input into computers. It might involve the use of a specialist data entry module of a package or program, a specific stand

Fig. 2.4. The Ferranti Market Research Terminal being used for entering the responses to a social survey questionnaire.

alone program, or a computer system editor. This type of data entry utilizes standard keyboards, although for operations such as entering numerical data for statistical analysis it may be assisted if the keyboard has a special separate pad of numeric keys.

Specialist data entry modules are found in many application packages and programs, such as the statistical analysis packages Minitab (see Chapter 4) and SPSS, and in specialist applications packages and programs, such as the survey analysis package SNAP (see Chapter 4) and the computer cartography package MICROMAP (see Chapter 5). Originally, these data entry modules were designed just to assist with entering data into the package/program itself. However, in an attempt to allow data transfer between packages/programs, standard data storage and transfer formats have been designated. Thus it is possible to make use of the very good data entry facilities in, say, SPSS/PC and then transfer the data into another package, such as Minitab, for analysis or display.

Specific stand alone data entry programs, because of the time they take to develop, are only warranted in certain special circumstances. Examples include, the program for entering the data collected by school children for the Domesday system (see Chapter 10) and the program for recording soil profile data described below. These were both developed to allow data entry at remote sites, with subsequent processing in a central computer after amalgamation of the data files.

These specialist data entry modules and specific stand alone programs have proved so useful because they improve both the speed and reliability of data entry. In such systems the data are normally entered in a structured format; for example, all the observations for the first variable are entered, followed by all the observations for the second variable etc. The computer can be used to perform some simple error checks, like examining if all the variables have the same number of observations and that only non-negative integer numerical data are entered. Most specialist data entry modules and programs also have facilities for correcting and editing data.

Computer system editors are specialist pieces of software which offer users a number of commands for creating and editing files. As such they may be used for general purpose data entry. The software is evoked by typing a command such as ED followed by the name of a file to be edited or created. A very wide range of commands are normally available to move around and to insert, delete and modify the contents of a file. A new version of an updated file can then be saved when all editing has been completed. Computer system editors do not normally have facilities for entering data in a structured format, or for error checking. They are, however, excellent for correcting data.

Using a soil profile recorder for data entry

Soil scientists were amongst the earliest users of quantitative methods and computers. A variety of attempts have been made in several countries to set up Soil Information Systems (SISs). These are computer-based systems which aim to assist in the storage, retrieval, manipulation and presentation of soil and land resource data (Rudeforth 1982). Although data collection and entry are amongst the most costly and time-consuming phases of setting up and maintaining a SIS, surprisingly, there have been few attempts to automate the process. One successful project described here makes use of a field portable microcomputer, in conjunction with specially written software, to assist in the process of direct manual data entry in the field.

In the Soil Survey of England and Wales soil classification scheme, the fundamental unit of soil data is the soil profile which is a vertical cut through the soil. A soil surveyor's job is to record information about the location and characteristics of a profile and associated environmental data. Traditionally, data are recorded in a surveyor's notebook using a proforma like that shown in Fig. 2.5. The rapidly increasing volume of information collected by surveyors, together with advances in information technology throughout the 1960s and 1970s, led to a realization that computers are ideal tools for improving the storage, analysis and presentation of soil data. Keyboard operators, not surveyors, have in the past been responsible for entering data into SISs. But a combination of bad handwriting, misinterpretation and the sheer volume of data make this operation both costly and error prone. This has stimulated the development of a portable device for entering soil profile information in the field.

The soil profile recorder comprises a field portable microcomputer, the Husky Hunter (Fig. 2.6), and a specially written computer program. The Husky Hunter has a number of characteristics which make it suitable for field use under most weather conditions (Clarke, Fisher and Ragg 1986; Fisher *et al.* 1987). The circuitry is all contained in a shockproof and weatherproof aluminium case measuring $216 \times 156 \times 32$ mm and weighing 1.1 kg. It is powered by four alkaline batteries, with a backup cell to protect programs and data when these are exhausted. The machine is available with enough memory to allow storage of up to 6–8 days work by a surveyor.

The software is written in BASIC and is based around a main menu which has six options. ENTER allows the input of environmental information and descriptions of soil horizons. Essentially, the input data are coded answers to the questions given in the proforma shown in Fig. 2.5. EDIT enables the surveyor to change data after input. SCAN facilitates the viewing of data. DELETE allows the surveyor to erase records from memory. TRANSMIT controls the automatic transfer of data to a host computer for processing (this may be another type of microcomputer or even a mainframe computer). Data may be transferred by direct link or via the public telephone network (see Chapter 9 for further details of how this may be achieved). The final option, FILE,

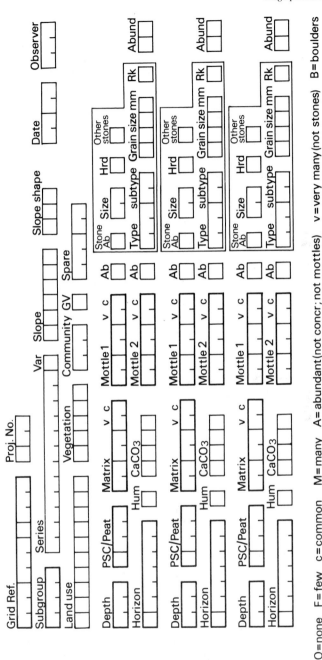

O=none F=few c=common M=many A=abundant(not concr.;not mottles) v=very many(not stones) B=boulders

X=extr. abund (not mottles) VS=very small S=small M=medium L=Large VL=very large

Fig. 2.5. The basic characteristics of the Soil Survey of England and Wales proforma for recording details of soil profiles.
Source: Clarke, Fisher and Ragg 1986.

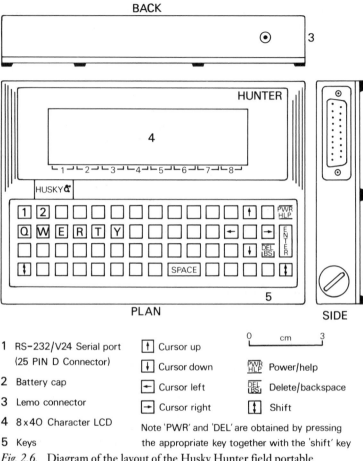

Fig. 2.6. Diagram of the layout of the Husky Hunter field portable
microcomputer which is the hardware element of the soil profile
recorder.

Source: Clarke, Fisher and Ragg 1986.

allows information to be stored in more permanent memory within the
microcomputer.

The soil profile recorder has already undergone field trials and is
proving to be a very useful aid to improving the efficiency and effectiveness of soil surveyors. Although the soil profile recorder described here
utilizes a Husky Hunter microcomputer, many other types of portable
computer could be used. Similarly, there are many applications outside
soil science to which this system could be adapted. Examples include,

the collection of responses to social survey questionnaires, topographic survey data and data about traffic flows.

CONCLUSION

This chapter has considered the process of data collection, one of the most crucial aspects of any geographical investigation. This is because it is often the most expensive and time-consuming phase of any investigation and because the methods of data collection have a great influence on the type of analysis which can subsequently be undertaken. In this chapter a basic classification of geographical data has been presented along with some examples of how data loggers linked to computers can assist in increasing the efficiency, improving the quality and reducing the cost of data collection. Computers have made a major contribution to the process of data collection in the past few years. This has been possible largely as a result of the advent of microprocessors and the concomitant reduction in the price of computer hardware. The signs are that this is a trend set to continue into the future. After the data collection process has been completed geographers proceed on to data management and analysis and these are the next chapters in this book.

FURTHER READING

Armstrong, A. C., Whalley, W. B. (1985) An introduction to data logging. *British Geomorphological Research Group Technical Bulletin* **34**. (A good readable introduction to data logging in physical geography.)

Shaw, G., Wheeler, D. (1985) *Statistical Techniques in Geographical Analysis*. Wiley, pp. 22–41. (A basic guide to geographical data sources, the principles of data collection and data handling.)

Persand, K. C., Virden, R. (1984) Data-logging with microcomputers. In **Ireland, C. R., Long, S. P.** (eds) *Microcomputers in Biology: a Practical Approach*. IRL Press, pp. 43–65. (A rather technical but useful introduction to data logging.)

Unwin D. J. (1981) *Introductory Spatial Analysis*. Methuen. (Good on classification of geographical data types.)

Geographical data management

This chapter is concerned with the storage and management of geographical data, an area in which some of the most significant recent developments in geographical computing have taken place. These developments have occurred for a number of reasons. All large collections of data which are accessed by more than one user require rules and regulations to maintain and manage them so that they remain in a usable state. The rapid increase in the volume of data, together with the development of high quality commercial software to assist with the organization and management of data, have also been significant factors. Additionally important is the fact that large data sets require efficient data storage structures, to ensure that they occupy the minimum amount of storage and that they can be effectively searched and analysed.

The basic concepts and structures which have been developed for storing and managing geographical data will be considered first. There is then a discussion of data base management systems including brief descriptions of ViewStore and the industry standard dBASE. Finally, an example of the application of data bases in geography is presented.

DATA BASE CONCEPTS AND STRUCTURES

Collections of related data on computers are organized into *files*. An individual file comprises a collection of *records* which may be thought of as individual cases, for example, one soil horizon or one questionnaire.

Each observation within a record is termed a *field*. In the example shown in Fig. 3.1 the first record in the file consists of data about the soil profile BAHD5. Stagnopodzol, Eg, 8, 45 etc. are all fields within this record.

Data files are classified into two types depending on the way in which computers store and subsequently access the data. The first type, *sequential access files*, have records organized in a linear sequence. To read any particular record (or field) it is necessary to start from the beginning of the file and read all records until the required record is located. Sequential access files are easy to set up and organize and they can easily cope with variable length records. On the other hand, this system of file organization has a number of disadvantages. These include difficulties of locating particular records, difficulties of inserting new records and the fact that the access time for records at the end of large files can be great. The second type, *direct access files*, are files in which access to an individual record (or field) is not related to its position in a file. This system of file organization allows easy updating of individual records. However, direct access files tend to take up more storage space, because unless the records are all the same length special steps need to be taken (for example, markers may be added to indicate the end of variable length records) and they are more difficult to maintain. Further discussion of file types may be found in Stephenson and Stephenson (1984).

To assist users, computer systems maintain a *directory* (also called a catalogue) of file names and ancillary information for each user. This enables users to organize their work, so that all the files for each application are located together, and to find out information about files, such as the size, date of creation and type.

Profile Code	Soil type	Horizon	%Loss-on-ignition	%Sand	%Silt	%Clay
BAHD5	Stagnopodzol	Eg	8	45	43	12
BAHD5	Stagnopodzol	Bs1	4	36	55	9
BAHD5	Stagnopodzol	Bs2	3	48	42	10
GA34	Stagnogley	Eag	7	30	55	15
GA34	Stagnogley	Bsg	6	32	50	18
GA34	Stagnogley	BG	6	31	52	17
EX113	Brown Soil	A	3	42	44	14
EX113	Brown Soil	B1	2	44	40	16
EX113	Brown Soil	B2	1	54	36	10

Fig. 3.1. A simple data base containing soil data for three soil profiles. Note that the text above the line is for annotation purposes only and is not part of the data base.

In applications which utilize limited quantities of data or require little data integration, only relatively simple concepts of data organization and management are required. Increasingly, however, it is necessary to make use of sophisticated models and techniques of data organization and management. It has become customary to refer to a large organized collection of data as a *data base*.

The process of centralizing and organizing data to form a data base offers six major advantages:

1 It forces the users of data to agree and enforce standards for data storage and exchange.
2 The change-over from a storage system, such as a filing cabinet, to a computer disk-based storage system reduces the amount of storage space required.
3 Data bases can be quickly and easily accessed in a range of different ways.
4 Selected parts of data bases can easily be passed to programs, for statistical analysis or graphical summary, or for incorporation into a word-processed report.
5 Data security can easily be maintained by allowing users access to only selected parts of a data base.
6 Much of the drudgery of manually updating and searching files can be reduced by using a computer.

The simplest model for a data base is the so-called *flat file*. This can be likened to the traditional card index of, say, map loans from a map library collection. Each record (card) in such a data base (card box) might contain fields consisting of the sheet number and title, the scale and the type of map (topographic, land use, soils etc.). In a flat file each record must have the same structure (same number of fields etc.) and the records in one file cannot be easily related to the records in another file. The data in a file are easier to search and manipulate if they are organized using some form of *index*. In a data base of map loans from a map library, for example, the records may be indexed by the mapping agency and sheet number or alphabetically by the borrower's name to give an *indexed flat file*.

The three basic models or structuring systems available for more complex data base applications are the hierarchical model, the network model and the relational model. A number of other less commonly used models have been identified, some of which are hybrids between the basic types (Bonczek, Holsapple and Whiston 1984). Although many types of data can be structured using several different data models, some models are more suitable than others for certain applications. The selection of the data model depends primarily on the characteristics of the data and the required uses of the data base.

In the *hierarchical model* each record can have several links to lower elements in a hierarchy, but only one link to a higher record (Fig. 3.2). Higher order records are usually called 'parents' and related lower order records are called 'children'. The highest record is called the 'root'. The postcode system of the United Kingdom lends itself to the hierarchical model. The root of the data base is the United Kingdom, which is divided into a series of 120 postcode areas. Each postcode area is divided into postcode districts, which in turn are divided into postcode sectors. These are divided into the smallest geographical unit, the unit postcode. There are about 1.5 million unit postcodes in the UK (see DoE 1987 for more details of the scheme). Further examples of data sets which could be suitably organized using a hierarchical model include plant, animal and soil taxonomy systems, and remote sensing images where pixels are classified into areas at several levels of aggregation. Hierarchical structuring is useful where data have one-to-many relationships and where only up-down searching is required. However, a large number of geographical data base applications require many-to-many relationships and necessitate links between records at the same level. Hierarchical data bases are generally the easiest data bases to create and maintain.

The *network model* is similar in some respects to the hierarchical model. However, a network data base is more general because a record can have more than one parent (Fig. 3.3). It is thus possible to represent many-to-many relationships. For example, a library data base will need to allow for the fact that a book may have only one borrower, but that a

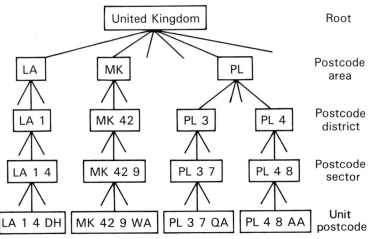

Fig. 3.2. Part of a hierarchical model data base of the UK postcode system shown in diagrammatic form.

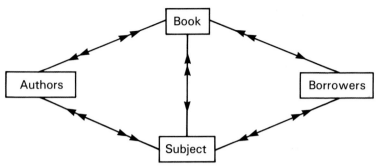

Fig. 3.3. Part of a network model data base of a library system shown in diagrammatic form. The arrows show the number of possible relationships between the records in the data base.

borrower may have more than one book. Similarly, an author may have books on many subjects and a subject may have books written by many authors. The network model may also be used, for example, to store the boundaries of areas in a spatial data base. Depending on the type of aggregation (travel to work area, television viewing area, local authority district etc.), the base units – which might be wards – need to be aggregated in different ways. Pointers in a network data base could be maintained to show how the wards are aggregated to give the various spatial units. Network data bases are usually the most difficult to create and understand. Moreover, because the computer has to maintain a list of pointers which link records together there can be large storage overheads associated with them.

In the *relational model,* data are organized in the form of a series of two-dimensional tables each of which contains one type of record (Fig. 3.4). The rows of the tables correspond to records and the columns correspond to fields of the records. In relational data bases, such tables (relations) are linked together by common field keys. Figure 3.4 shows how data collected on a postcode sector basis, as part of a shopping survey (Table 1), may be linked with earlier data collected on a ward basis during a national population census (Table 2). The key fields in this example are postcode sector and ward. If we wish to determine the importance of car ownership, a simple ratio of the average number of shopping trips per car will be of assistance. However, because the data are available on different spatial bases, a third table is required which shows how to link the two data sets together (Table 3). This shows that when the postcode sectors are overlaid on top of the wards, 50 per cent of the ward DAAD and 100 per cent of the ward DAAE make up the area of PL73 and that 75 per cent of DAAF and 50 per cent of DAAA make up the area of PL74 etc. Using these values it is possible to

Table 1

Postcode sector	Number of shopping trips/year
PL73	4000
PL74	8000
PL75	8000
PL76	1000

Table 2

Ward	Number of cars
DAAD	200
DAAE	400
DAAF	500
DAAG	500

Table 3

Postcode sector	Ward	Percentage
PL73	DAAD	50
PL73	DAAE	100
PL74	DAAF	75
PL74	DAAD	50
PL75	DAAF	25
PL75	DAAG	43
PL76	DAAG	57

Number of shopping trips/car

$PL73 = 4000 / [(200/0.50) + (400/1.00)] = 5.0$
$PL74 = 8000 / [(500/0.75) + (200/0.50)] = 7.5$
$PL75 = 8000 / [(500/0.25) + (500/0.43)] = 1.9$
$PL76 = 1000 / [(500/0.25)] \qquad\quad = 1.4$

Fig. 3.4. A simple example of a relational data base for integrating shopping survey and car availability data. The data in Tables 1 and 2 are linked together by key fields to give Table 3 and data on shopping trips/car.

calculate the required ratios as is shown in Fig. 3.4. The variation in the number of shopping trips per car is explained partly by the number of cars per household (PL76, for example, is a very prosperous area), but the results also show that the distance from the areas to the shops is also important (PL74, for example, is close to a number of supermarkets, making it convenient to visit shops frequently). This type of data structure can also be used to set up spatial data bases in which lines are linked together to represent polygons (areal units). In this way, common boundaries between polygons need only be digitized once (Burrough 1986).

Relational data bases are very flexible and are reasonably easy to understand and set up, especially because they are based on sequential access files. Insertions, updates and deletions can be carried out without undue difficulty. Like the network model, the relational model allows for many-to-many relationships.

DATA BASE MANAGEMENT SYSTEMS

A *Data Base Management System* (DBMS) is a computer program for creating, maintaining and accessing a data base. One of the many functions of a DBMS is to isolate end users from the complexities of computer systems and provide them with an easy to use management and analysis system. Thus, although crucial in determining the usefulness of a data base for some applications, the data storage structure is often transparent to the end user (most geographers fall into this category).

Data bases have been used in a wide variety of geographical investigations. MacDougall (1983) used DBMSs in two projects. In the first, information about the physical characteristics (maximum height, maximum spread, hardiness etc.) of trees was used to select suitable plants for use in urban building programmes. In the second, information about land use type, accessibility and soil capability was used for regional analysis and landscape assessment. Daker (1987) also describes a data base containing details of the urban structure of a small town, that has been used in teaching urban geography. Examples of data bases of map catalogue, bibliographic and census data are also described below.

Most commercially available DBMSs deal with numerical or alphanumerical data. A small number of packages are available which allow pictures and maps to be linked with alphanumerical data and stored in data bases. Two of the most widely known packages are Filevision and Mapvision for the Apple Macintosh. In these packages graphic images (maps, graphs, pictures etc.) can be stored, retrieved and edited in much the same way as alphanumerical data. Filevision is reviewed by Stobie (1985).

The principle features of a DBMS important for geographical users relate to the creation of a data base, the extraction of data from a data base, the transfer of data to and from other software packages and data protection. These will be discussed in turn below. General reviews of some commercial DBMSs may be found in Bond (1984) and Lang (1985a).

Creating a data base

It almost goes without saying that it is essential to plan the structure and content of a data base very carefully before any data are entered. The structure and content of any data base is influenced by a number of factors. The type of data influences such things as whether the data are to be stored in code or in full, the length of each field and the number of records. The type of DBMS used to create a data base affects, for example, how much data can be entered and what length records can be entered or viewed. The purposes for which the data base is required will influence, for example, which data fields will be indexed (pre-sorted into ascending or descending order to allow rapid retrieval) and how much of each field will be displayed on the screen.

Data base management systems usually have a utility (a program routine) to enable users to set up new data bases (in the ViewStore DBMS it is called SETUP and in dBASE III it is called CREATE). The set up utility prompts users, on a question and answer basis, for the necessary information to create a new data base. Once the structure has been defined the data base can be built up by entering records. Data can be entered either via the keyboard or data files can be transferred from other packages (see below). The commands which define the format of a data base are normally saved in a format file on disk. A second file which contains the actual data in the data base will also be created on disk. Once a data base has been created a variety of commands are available for inserting, changing and deleting records.

Extracting data from a data base

Most data base management systems allow data to be displayed in more than one format. The exact details of the display are defined in a format file, but basically two main types of display are possible – the spread-sheet (or table) display and the card display. The spreadsheet display shows a single record in abbreviated form on each line of the screen (Fig. 3.5). The card display shows the title and contents of every field in a single record. Each record is set out as though written on a paper reference card (Fig. 3.6). In both types of display the screen is best thought of as a window which shows only a portion of a much larger sheet of electronic paper. It is possible to move the window around the paper using cursor keys. When a cursor key is pressed the DBMS automatically updates the screen display by reading the appropriate records from the data base file.

One of the most widely used facilities of DBMSs enables users to identify specific categories of records in data bases. For example, in the case of a bibliographic data base, it is possible to select a single refer-ence, by typing in say the author of a book (Fig. 3.6). Alternatively, all references by a given author, or all references on a particular subject may be selected. The selection criteria can be based on any value in any field in the data base. It is also possible to build up very complex

```
L Space 35    Indexed by AUTHOR
                          References
Enter the name of the Author(s) e.g. Smith, J.A.

AUTHOR.............YEAR.TITLE......................JNL.......................
Anderson, T.D.       1985 Geopolitics of the Caribbean *************************
Bell, J.P. & McCul 1969 Soil moisture estimation by  J. Hydrol., 7, 415-433.
Bigelow, G.E.        1984 Simulation of pebble abrasio Earth Surface Processes an
Bird, E.C.F.         1984 Coasts: an introduction to c 3rd Edition. Blackwell, Ox
Bleasdale, A.        1957 Hydrology and British rainfa Meteorol. Mag.,86, 206-210
Bleasdale, A.        1959 The measurement of rainfall. Weather, 14, 12-18.
Bleasdale, A. & Do 1952 Storm over Exmoor on August  Meteorol. Mag., 81, 353-36
Bleasdale, A.        1963 The distribution of exceptio J. Inst. Water Eng. Sci.,
Bowden, K.           1953 Storm surges in the North Se Weather, 8, (3), 82.
British Standards    1964 The measurement of liquid fl B.S. 3680 Parts 1, 2, 3, 4
Brunn, P. & Gerrit 1959 Natural by-passing of sand a Proc. Am. Soc. Civ. Engrs.
Bunting, B.T.        1961 The role of seepage moisture Am. J. Sci., 259, 503-518.
Burton, I. & Kates 1964 The perception of natural ha Natural Resources Journal,
Burton, I., Kates, 1978 The environment as hazard.  Oxford University Press.
Burton, J.D. & Lis 1976 Estuarine chemistry.         Academic Press, London.
Calkins, D. & Dunn 1970 A salt tracing method for me J. Hydrol., 11, 379-392.
Carr, A.P.           1962 Cartographic record and hist Geography, 47, 135-144.
Carr, A.P.           1965 Shingle spit and river mouth Trans. Inst. Br. Geogr., 3
Chambers, F.M. & P 1985 Palaeoecology of Alnus (alde New Phytol., 101, (?), 333
Chatterton, J.B.     1977 Nottingham flood warning sch Middlesex Polytechnic Floo
```

Fig. 3.5. A spreadsheet display of part of a bibliographic data base developed using ViewStore on an Acorn BBC microcomputer (see also Fig. 3.6).

selection conditions by using relational operators (such as '>', '<', and '=') and wild cards. The latter are special characters which can be used to select similar fields. In ViewStore the character '?' is used to match any single character and the character '*' is used to match any number of any characters (including none). In the case of the bibliographic data base the search criterion 197* can be used to match all references published in the 1970s.

In addition to this form of search facility, there are some systems which have facilities for geographical selection of records. The map retrieval system, described by Pearson and Sprunt (1987), makes use of such an interface. The system, which is based on dBMAN (a dBASE clone) on an Atari microcomputer, allows users to select records containing information about maps in a map catalogue by pointing to a specific area on a small scale base map.

A second widely used facility of DBMSs is sorting. This entails sorting records into either ascending or descending order on the basis of one or more fields. The sorted records, which might have been previously selected from the data base can then be displayed on the screen, printed or stored in a file for future use.

Once the required data have been extracted from a data base and displayed on the screen it is often desirable to produce a hard (paper) copy of the data. Most DBMSs have facilities for printing out data in a

```
L Space 35      Indexed by AUTHOR
                                References
Spare slot

    AUTHOR Bird, E.C.F.

      YEAR 1984

      TITLE Coasts: an introduction to coastal geomorphology.

        JNL 3rd Edition. Blackwell, Oxford.

    LIBREF 551.447BIR

KEYWORDS Coastal

      LIST CZM

        MAC N

       TEXT T

      SPARE ▇▇▇
```

Fig. 3.6. A card display of one of the records from the data base shown in Fig. 3.5.

range of formats including the card display and the spreadsheet display. Some also provide commands to print data in more complex output styles such as sticky label and table format.

Transferring data between software packages

All good DBMSs should have facilities for transferring data between various software packages. This may involve transferring data from, say, dBASE III into the spreadsheet package Lotus 1–2–3, or passing INFO data files to the statistical analysis package SPSS. To make maximum use of the wide variety of geographical applications software available for computers, it is essential that data can be transferred between programs. Unfortunately this is only available on an *ad hoc* basis at present.

Data protection

There are two main aspects to the protection of data in data bases. The first relates to protecting data against the threats of accidental loss or unauthorized examination, copying or destruction. The second relates to protecting the rights of individuals by giving them access to personal data and by regulating the activities of the holders and users of such data.

The best way to protect data against accidental loss is to formulate a disaster recovery plan. This usually involves making backup copies of critical disks and documentation. Before a disaster recovery plan can be considered reliable, it must be given a full scale test.

The protection of data against unauthorized threats is rather more difficult (Shain 1986). Basically it involves controlling access to computer systems and making data unintelligible. There are two basic aspects to access control: authentication and authorization. Authentication refers to the computer system identifying users by requesting something about them (fingerprints, signature etc.) something known by them (passwords, personal information etc.) or something possessed by them (identification cards, keys etc.). Authorization refers to the hardware and software techniques which restrict what users may or may not do once they have entered a system. Since no access control systems are totally foolproof, data may be further protected by storing and transmitting them in coded forms. This makes it unintelligible to anyone without the key to the code.

In the United Kingdom, the Data Protection Act 1984 is designed to protect the rights of individuals. This requires all those who hold and use automatically processed data to register their activities and to comply with the eight principles of the law (Sterling 1986). Many other countries have similar legislation.

ViewStore

ViewStore is a general purpose data base management system for Acorn microcomputers. It is supplied by Acornsoft and it is compatible with the word processing package View, the spreadsheet package ViewSheet and the rest of the View range. ViewStore is reviewed in Lang (1985b) and Someren (1986).

ViewStore has some of the features of flat file packages, that is, each record must have the same number of fields and two dissimilar sets of records cannot be related. It also has many of the attributes of considerably more sophisticated systems. A number of utilities are provided within ViewStore to create, maintain and access data bases. u.SETUP is used to create new data bases and reports. Users are prompted for information about the name of the data base, the number of records, the size of fields etc. u.INDEX is used to create an index for any field which can then be employed to order records quickly. u.SELECT is used to select subsets of records from a data base. u. REPORT is useful for printing extracted data and allows a range of formats, including tables and text, to be combined in quite sophisticated ways. u.LABEL is used to print data in label format on either the screen or on a printer. This utility can be used to print sticky address labels. u.CONVERT enables existing data bases to be converted into a new format by reordering fields, changing the size of fields etc. u.LINK allows files to be transferred to and from other packages in the View range. A command called IMPORT, although not strictly a utility, is provided for converting files from other data bases (such as dBASE II) into ViewStore format files. u.MACRO allows data extracted from a ViewStore file to be incorporated into View text files such as standard letters.

Although ViewStore is ideal for many simple geographical data base applications, it does have some limitations of which the restriction to one range of microcomputer is most severe. Whilst the documentation is good, ViewStore does not have any on-line help facilities and it is not menu driven (that is, it does not have facilities for presenting a list (menu) of possible options on screen from which the user can choose). At a more specific level, it is not possible to open more than one data file at the same time, nor can ViewStore handle several sets of records of dissimilar structure.

dBASE

dBASE II and dBASE III are relational data base management systems developed by Ashton-Tate, which are available for a wide range of popular microcomputers and have been adopted as the *de facto* standard DBMS for microcomputers. Both dBASE II and III have a flexible and powerful enquiry programming language for integrating large data base files. They have a reputation for being very powerful but unfriendly, because they are difficult to learn to use. As a consequence of this, many 'front end' application programs (that is, programs which provide an alternative, usually easier to use, program and or command structure for the user) have been written, to configure the packages to handle specific tasks. Many books and articles have been written about their use (see for example Byers 1983; Perry and McJunkins 1985).

Both dBASE II and III have a very wide range of commands for editing, retrieving, manipulating and printing data (including all those described earlier in this chapter). The built-in programming language makes them potentially very flexible. Files can be processed both interactively and in batch mode using the command DO 'filename' (where the name of a file containing a series of commands is substituted for 'filename'). Command loops and procedures (similar to FOR-NEXT loops and subroutines in BASIC – see Chapter 12) can also be established. It is possible to access other packages, such as the word processing package WordStar (see Chapter 8), from within dBASE II and III.

In conclusion, dBASE II and III are extremely powerful data base management systems. Many of the limitations of dBASE II, such as the restricted file and record size and the difficulty of use, have been rectified in dBASE III. Both are extremely well documented and tested.

Although most users believe them to be good value for money, they may prove prohibitively expensive for some applications. In response to this, a number of low cost dBASE II emulators have been marketed, such as Practibase which has all the features of dBASE II and others besides (Lewis 1986). For example, it has extended file and record size and it is menu driven.

CREATING AND USING A DATA BASE OF CENSUS DATA

The decennial Census of Population of Great Britain provides an unrivalled data source for geographers, planners and other social scientists. The Census represents the most comprehensive source of data readily available in Britain at a small scale, on demographic, social and economic conditions. It has been widely used to provide an indication of the social and economic problems which need to be tackled and where resources need to be directed.

The Census was last taken on 5 April 1981. A questionnaire, to each of the 54 million individuals in Britain, yielded eight to fifteen primary answers. Added to these, there were a further six answers relating to each household. This information, which amounts to 4,400 statistical counts for each of the 130,000 base areas in Britain, is summarized in published tables in paper and computer format. In England and Wales the base areas are Enumeration Districts (EDs) which typically comprise about 200 households. In Scotland the postcode system is preferred, with unit postcodes comprising about 14 households as the base units (Rhind 1983).

The small area statistics (EDs, wards and districts: unit postcodes, postcode sectors and postcode districts) from the 1981 Census, may be accessed using a specially written software package called the Small Area Statistics PACkage (SASPAC). This allows users to extract subsets of the data (certain variables for certain areas) and to perform some simple statistical calculations.

Unfortunately, because of its size and cost and because it was designed to run on a variety of minicomputers and mainframe computers, the process of extracting small area census data using SASPAC suffers from a number of difficulties (though it still remains a milestone in providing widespread access to a digital data base). The operating procedures of the computers on which the small area statistics and SASPAC are available mean that SASPAC can rarely be run interactively. Turnaround times for jobs are, therefore, often considerable. Additionally, because of the general purpose nature of the software, only a limited number of search and analysis functions are available.

Furthermore, novice computer users often have difficulty in using the package due to its complexities.

To combat these problems, a subset of the small area statistics for Plymouth District has been extracted and established as a data base for research and teaching in geography. The data base has been set up on an Acorn BBC microcomputer using ViewStore, though it must be emphasized that many other combinations of hardware and software could have been used to achieve similar results. ViewStore was chosen since it is relatively cheap, is simple to use and is available on the Acorn BBC microcomputer (which is common in educational establishments in Britain). A similar exercise has also been undertaken by Stewart (1987) who created a data base of information for Thamesdown, Wiltshire, UK.

The SETUP utility in ViewStore was used to create the data base. This is an easy-to-use interactive module which requests information, such as the number and size of records and fields, the order in which the fields are to be displayed on the screen and the name of each field. The requisite data were then extracted from the small area statistics using SASPAC (held on Plymouth Polytechnic's minicomputer) and transferred into the data base using the IMPORT command in ViewStore.

The Plymouth Census Data Base comprises 522 records (one for each of the EDs in Plymouth District) and 19 fields (variables). The records all have the same number of fields, but the size of the fields varies in length, because of differences in the size of data items which they hold. Some of the fields contain alphanumerical and others contain only numerical data. A range of geographical, demographic, economic and social variables was chosen to illustrate the character of the District (a selection of twelve are shown in Figs 3.7 and 3.8).

The data can be displayed in both card and spreadsheet format and can be indexed using one of nine specially created indexes. For example, the data in Fig. 3.7 are displayed in card format and are indexed on the field ED (this index sorts the records into alphanumerical order).

The most powerful feature of the system and the one which makes it so useful, is the ability to select specific categories of records. For example, it is possible to select all EDs in a given ward, or all EDs with greater than 20 per cent of the households without a car. The selection criteria GAAA* (where '*' is used to match any number of characters) will match all the EDs in ward GAAA (Budshead). This retrieves 21 records (EDs) from the total of 522 in the data base. Multiple selection criteria can also be used, for example, to build up a picture of the social geography of the area. A study of the mobility of the elderly in the dockyard area of Plymouth (Keyham ward), may require data that highlight areas where there are large numbers of households containing one or more pensioner and no car. These can easily be obtained from the data base using the select criteria WARD = KEYHAM AND

```
L Space 20      Indexed by ED
              Plymouth Census E.D. data. Values/1000 households
Enumeration district

ED......EAST.NORTH.WARD.......OLD.YOUNG..PENS.SINGL.THREE..MIGS.SOCHI.SOCLO.SER
```

ED	EAST	NORTH	WARD	OLD	YOUNG	PENS	SINGL	THREE	MIGS	SOCHI	SOCLO	SER
6AAA11	02466	00602	BUDSHEAD	205	220	309	200	46	73	0	286	
6AAA12	02469	00600	BUDSHEAD	115	279	158	110	72	86	0	333	
6AAA13	02465	00600	BUDSHEAD	208	241	275	147	31	87	0	500	
6AAA14	02469	00598	BUDSHEAD	224	188	320	202	40	131	53	158	
6AAA15	02471	00596	BUDSHEAD	238	254	272	224	56	185	0	227	
6AAA16	02468	00603	BUDSHEAD	149	256	183	153	106	76	48	190	
6AAA17	02475	00597	BUDSHEAD	221	216	301	186	56	45	111	222	
6AAA18	02474	00596	BUDSHEAD	328	148	464	374	21	122	0	250	
6AAA19	02476	00595	BUDSHEAD	147	250	210	130	93	54	143	48	
6AAA20	02479	00593	BUDSHEAD	186	225	240	147	86	56	0	250	
6AAA21	02481	00595	BUDSHEAD	253	211	358	251	89	48	105	158	1
6AAB01	02483	00573	COMPTON	401	149	431	249	48	107	526	0	
6AAB02	02485	00576	COMPTON	222	191	269	118	69	70	176	59	1
6AAB03	02487	00575	COMPTON	69	260	60	53	82	177	471	0	2
6AAB04	02488	00572	COMPTON	137	280	232	232	72	93	167	0	1

Fig. 3.7. A spreadsheet display of part of the Plymouth Census Data Base developed using ViewStore on an Acorn BBC microcomputer. The column headings are as follows: ED (enumeration district) SASPAC code, EAST (NGR easting), NORTH (NGR northing), WARD, OLD (number of male persons of retirement age), YOUNG (number of children aged 9–15), PENS (number of households with one or more person of retirement age), SINGL (number of unmarried adults), THREE (number of households with three or more dependent children), MIGS (number of households containing migrants – persons who have moved within the last twelve months), SOCHI (number of households whose head is employed in SEGs 1–4), SOCLO (number of households whose head is employed in SEGs 10 and 11), SER (number of households whose head is in the armed forces).

(OLD > 250 AND NOCAR > 350). This retrieves 7 records from the data base.

Figure 3.8 shows a list of all EDs whose centroid (centre point) lies within 0.5 km of the Department of Geographical Sciences, at Plymouth Polytechnic (National Grid Reference 02481 00550). The search was carried out on the fields EAST and NORTH, which are the National Grid References of the ED centroids (as given in the Census). The data base was searched using the select criteria EAST >=2476 AND EAST <2486 AND NORTH >=0545 AND NORTH <0555. Using this selection criteria 16 records were selected.

The Plymouth Census Data Base provides a useful example of the potential utility of data bases in geography. Using the data base it is possible to locate information quickly and to test geographical hypotheses about the various facets of the social and economic geography of the area. The results can be printed out in a variety of formats and it is possible to transfer the data and results to one of the other packages in

```
=>UTILITY SELECT
SELECT
List or create select file (L,F)? L
Screen or Printer (S,P)? S
Select criteria? EAST>=2476 AND EAST<2486 AND NORTH>=0545 AND NORTH<0555
Select criteria?
```

ED	EAST	NORTH	WARD	OLD	YOUNG	PENS	SINGL	THREE	MIGS	SCLH	SCLL
GAAC18	02481	00553	DRAKE	307	101	316	527	13	204	143	214
GAAC19	02480	00553	DRAKE	288	112	301	450	36	255	167	111
GAAC24	02478	00552	DRAKE	283	165	426	327	48	159	143	71
GAAC25	02477	00551	DRAKE	228	129	268	312	23	130	83	250
GAAC26	02480	00550	DRAKE	305	121	315	391	18	179	182	182
GAAC27	02484	00554	DRAKE	178	117	200	357	39	276	214	71
GAAC28	02484	00553	DRAKE	195	84	228	421	19	310	118	235
GAAC29	02483	00552	DRAKE	210	160	248	314	60	151	71	143
GAAC30	02482	00551	DRAKE	278	177	333	338	38	213	143	143
GAAC31	02485	00550	DRAKE	247	111	274	395	20	265	125	63
GAAC32	02483	00548	DRAKE	289	131	352	402	18	114	63	125
GAAC33	02483	00549	DRAKE	204	136	294	371	22	240	77	77
GAAC34	02482	00550	DRAKE	284	75	321	231	0	246	333	333
GAAR11	02476	00550	ST.PETER	505	147	561	433	16	125	0	0
GAAU02	02476	00546	SUTTON	390	134	348	511	21	82	0	200
GAAU13	02484	00546	SUTTON	318	106	393	483	35	200	67	67

```
16 records selected out of 522
```

Fig. 3.8. Enumeration Districts selected from the Plymouth Census Data Base which are within 0.5 km of NGR 2481 0550.

the View range, for incorporation into a word-processed report (View), for statistical analysis (ViewSheet), or for graphical presentation (View-Plot). Although the system is very useful it does have its limitations. The size of the data base is limited by the size of the Acorn BBC micro-computer memory and disk storage. The speed of the microcomputer also means that some searches can take a minute to process. There are also limitations posed by the availability of the hardware and by the restricted range of data in the small area statistics. Nevertheless, the data base has been successfully used for over a year in both geographical research and teaching.

CONCLUSION

The recent rapid proliferation of data about the cultural and natural environment has created a great demand for data storage and management tools. The large number of commercial packages available for many types of computers can easily and readily be used by geographers. This chapter has described the basic characteristics of data bases and the structures which have been used to organize data. Examples have been presented which illustrate how data base management systems can be used to create, organize and extract data from a data base. As geographers become increasingly aware of what commercial data storage and management tools are available and how easily they can be used, the number of applications of data base management systems is sure to rise. Already the signs are that this is becoming one of the fastest growing areas of computer use in geography and related disciplines.

FURTHER READING

Burrough, P. A. (1986) *Principles of Geographical Information Systems for Land Resources Assessment. Monographs on Soil and Resources Survey* **12**. Clarendon Press. (A good introduction to the use of data bases in a geographical context.)

Date, C. J. (1981) *An Introduction to Data Base Systems* 3rd edn. Addison-Wesley. (The data base bible, but rather complex for beginners.)

Lang, K. (1985a) *Business Computing: the Survival Game. Personal Computer World Special*, pp. 63–66. (A readable introduction and review of the main UK microcomputer DBMSs.)

Laurie, P. (1983) *Databases: How to Manage Information on your Micro.*

Chapman and Hall/Methuen. (An introduction to data bases on microcomputers, although rather poorly structured.)

Rhind, D. W. (1983) (ed) *A Census User's Handbook.* Methuen. (An excellent introduction to the census and techniques of geographical analysis.)

Statistical analysis

Computers have traditionally been associated with the manipulation and storage of numbers. Indeed, many of the early machines were designed specifically for such operations. Although now employed for many other purposes as well, they are still widely used by scientists and social scientists to undertake a variety of statistical analyses. The next sections of this chapter will consider the basic characteristics of the main types of statistical software currently available, namely subprograms/subroutines, programs and packages. Two examples of statistical analysis using a spreadsheet (ViewSheet) and a general purpose statistical analysis package (Minitab) are presented. A number of statistical analysis systems are discussed in the text, although statistical techniques themselves are not considered in detail. Readers unfamiliar with statistical techniques are directed to one of the standard geographical texts such as Silk (1979) and Johnston (1980).

In the 1960s and 1970s, following the quantitative and theoretical revolution, and as a central component of the positivist paradigm, statistical analysis was given great prominence in geography. During the 1960s it was considered to be in the vanguard of the subject and by the beginning of the 1970s it reached a new level of maturity, pervading the whole discipline and becoming one of the core areas (Bennett and Wrigley 1981). In the late 1970s and the 1980s, however, many workers began to question the philosophical underpinnings and relevance of the positivist paradigm and thus the value of statistical analysis. A number of geographers argued for an increase in the attention devoted to alternative paradigms, such as humanism and structuralism (Johnston 1987).

The methodologies suggested by these paradigms should not, however, be seen as mutually exclusive. It would be foolish and arrogant to argue either that statistics can be used to solve all geographical problems or that they cannot be used for any. Statistical analysis remains a very important technique which can be widely employed in a great many areas of geography.

Statistics can be used for many purposes. They can be used to determine how much data should be collected so that an accurate understanding of phenomena can be developed. This is a type of cost benefit analysis, where the extra benefit gained from collecting more data is offset against the extra cost involved in collecting them. Statistics can be used to describe, classify and summarize data. Thus a complex collection of data can be reduced to a few numbers which represent the key elements. Using statistics it is possible to place error limits on data. For example, an average value of 12 per cent moisture content for ten soil samples, is of only limited value unless the validity and reliability of this value are also known. Statistics allow us to make inferences about populations using a sample of a small amount of data. For example, if data were collected about the spectral reflectance of tree canopies in selected areas, statistical inference could be used to draw conclusions about the spectral reflectance of tree canopies in general. Lastly, statistics enable us to be more rigorous in our work and to avoid self-deception.

The development of statistical methods in geography, not surprisingly, has been closely linked to the development of computers. The ability to store large quantities of data which can be manipulated quickly in a range of different ways, has made the use of computers almost obligatory for any type of complex statistical analysis. Indeed, prior to the advent of computers, some statistical analyses were simply too time-consuming and complex to contemplate. Even now Openshaw (1987) argues that only by using the most powerful supercomputers can some types of urban and regional modelling be undertaken.

In addition to making a profound impact on geographical research, computers have also substantially altered the way statistics are taught to geographers. The speed and reliability of computers has allowed more complex statistical tests to be carried out on data sets of a larger and, therefore, more realistic size. As a result, more time has been made available to experiment with statistics. For example, using computers it is very easy to examine the effect of transforming variables in regression analysis. Perhaps more important, as far as geographers are concerned, is the fact that using computers it is possible to place more emphasis on the interpretation rather than the calculation of statistics.

For statistical analysis little additional hardware is required beyond the basic components of a computer. Although if statistical analysis is to be the main use of a computer it may be worthwhile ensuring that the hardware can process numerical data quickly, in large units (if individual numbers are processed in small units significant rounding errors can occur) and that it can handle a large range of numbers (at least 2×10^{-128} to 2×10^{127}). It may also be worthwhile checking that the computer keyboard has an appropriate selection of suitably displaced numerical and arithmetical function keys and that there is adequate disk storage available for holding data. In contrast, it is rather more important to give due consideration to the type of software used in an investigation. In particular, statistical analysis software must be easy to use, flexible, bug-free and computationally correct (see Weatherill and Curram 1984 for further discussion). It is, for example, very frustrating to enter 300 or so data items only to find that there is a bug in the program or that it gives results outside the permissible range.

Chambers (1980) suggests that there are three main types of statistical analysis software: (a) collections of subprograms or subroutines; (b) single programs; and (c) statistical systems or large package programs. These will be dealt with in turn below. General reviews of computer software may be found in Carpenter, Deloria and Morgenstein (1984), Fox (1984), Weatherill and Curram (1984), Gilchrist (1985), Rowe *et al.* (1985), O'Brien (1986), O'Keeffe and Klagge (1986) and Cable and Rowe (1987).

SUBPROGRAMS/SUBROUTINES

This approach to statistical computing involves the development of algorithms and their implementation as subprograms or subroutines. Individual subprograms/subroutines are then combined together to produce a complete program to carry out one or more tasks. This approach offers great flexibility since many combinations of algorithms are possible and tailor-made programs can be developed relatively quickly. Libraries of subprograms/subroutines can also be updated frequently as new algorithms are evolved. The main disadvantage to users is that programming skills are required in order to implement the subprograms/subroutines and link them with data. It is essential, therefore, when using subprograms/subroutines, that good documentation is available and that the programmer who is implementing them has some knowledge of how the subprograms/subroutines work and what they were designed to achieve.

The Numerical Algorithm Group (NAG), for example, has produced a library of subroutines written in FORTRAN, many of which may be used for statistical analysis. These have been widely used by geographers in the past. Cooke, Craven and Clarke (1982, 1985) have also

```
1000 REM *** EXCHANGE SORT ***
1010 REM DIMENSION A(N), NUMBER OF DATA=N
1020 REM LOCAL VARIABLES: I,J,J1,W
1030 FOR I = 1 TO N-1
1040 LET J1 = I
1050    FOR J = I+1 TO N
1060    IF A(J) >= A(J1) THEN 1080
1070    LET J1 = J
1080    NEXT J
1090 IF J1 = I THEN 1130
1100 LET W = A(I)
1110 LET A(I) = A(J1)
1120 LET A(J1) = W
1130 NEXT I
1140 REM DATA OUTPUT IN INCREASING ORDER IN A(N)
```

Fig. 4.1. BASIC subprogram for sorting data.

developed a series of subprograms, written in BASIC and Pascal, which may be implemented on most types of computer. Figure 4.1 is an example of one of the BASIC subprograms from Cooke *et al.* (1982), which can be used to sort a set of data into ascending order. The data to be sorted are stored in the array A(N) and the number of observations is stored in the variable N.

SINGLE PROGRAMS

Since the beginning of the 1960s geographers have been writing programs to assist with the calculation of statistics. These single programs are by definition stand alone and self-contained, although some have been grouped together to form program libraries (Unwin and Dawson 1985). They have the advantage that all the user needs to do is run the program, enter the data and read the output. The main disadvantages with single programs are that they are task-specific and that combining programs may be difficult.

Many of the recent statistics texts books, on quantitative aspects of geography and other sciences and social sciences, also contain a number of statistical analysis programs. The best and most comprehensive text of this nature is Cohen and Holliday (1982). This book is designed to be used as a statistics course text for social science students. Forty-six statistical tests are illustrated, each of which is accompanied by a BASIC computer program. Figure 4.2 shows the chi-square program (PROG-1O). Other texts which contain computer programs that may be used to solve geographical problems include Lee and Lee (1982), Bishop

```
 10 REM------------------------------------
 20 REM PROGRAM TO COMPUTE THE CHI-SQUARE
 30 REM ONE-SAMPLE CRITERION FOR
 40 REM DIFFERENCES BETWEEN OBSERVED AND
 50 REM EXPECTED FREQUECIES.  EXPECTED
 60 REM FREQUENCIES DIVIDED EQUALLY
 70 REM AMONG CATEGORIES
 80 REM     A=PRIMARY DATA MATRIX
 90 REM     N=NUMBER OF CATEGORIES
110 DIM A(100)
120 PRINT"CHI-SQUARE ONE-SAMPLE TEST"
130 PRINT"========================="
140 PRINT
150 REM------------------------------------
160 REM DATA INPUT
170 REM------------------------------------
180 PRINT"INPUT NO. OF CATEGORIES";
190 INPUT N
200 PRINT
210 PRINT"INPUT NO. OF FREQUENCIES IN"
220 PRINT"EACH CATEGORY."
230 PRINT"ERRORS IN INPUT CAN BE"
240 PRINT"CORRECTED AFTER ALL DATA"
250 PRINT"ENTERED"
260 PRINT
270 FOR I = 1 TO N
280 PRINT " CATEGORY ";I;" = ";
290 INPUT A(I)
300 NEXT I
310 REM------------------------------------
320 REM CORRECTION ROUTINE
330 REM------------------------------------
340 PRINT"ARE THE DATA CORRECT";
350 INPUT A$
360 IF A$="Y" THEN 490
370 PRINT
380 PRINT"WHICH CATEGORY CONTAINS"
390 PRINT"INCORRECT DATA";
400 INPUT C
410 PRINT"PRESENT VALUE IN CAT. ";C;" = ";A(C)
420 PRINT"WHAT IS THE CORRECT VALUE";
430 INPUT A(C)
440 GOTO 340
450 REM------------------------------------
460 REM COMPUTE EXPECTED FREQUENCIES
470 REM------------------------------------
480 LET S=0
490 FOR I = 1 TO N
500 LET S=S+A(I)
510 NEXT I
```

```
520 LET E=S/N
530 REM------------------------------
540 REM COMPUTE CHI-SQUARE
550 REM------------------------------
560 LET D=0
570 IF N=2 THEN 650
580 FOR I = 1 TO N
590 LET D=D+((A(I)-E)^2)/E
600 NEXT I
610 GOTO 680
620 REM------------------------------
630 REM YATES CORRECTION
640 REM------------------------------
650 FOR I = 1 TO N
660 LET D=D+((ABS(A(I)-E)-0.5)^2)/E
670 NEXT I
680 PRINT
690 PRINT
700 PRINT"------------------------------"
710 PRINT"    CHI-SQUARE= ";D
720 PRINT"    DEGREES OF FREEDOM= ";N-1
730 PRINT"------------------------------"
740 END
```

Fig. 4.2. BASIC program for chi-square.

(1983), Sharp and Sawden (1984), Ebdon (1985), Ottensmann (1985), Tennant-Smith (1985) and Williams (1985).

PACKAGES

Packages are really large complex programs or integrated collections of programs which have a special language for giving instructions. To make full use of a statistical analysis package it is necessary to be familiar with the language syntax and the strengths and weaknesses of the facilities available. Not surprisingly, packages tend to be more expensive than subprograms/subroutines and programs. Their large size also means that powerful, and therefore, expensive hardware is required to run them. On the other hand, they often have a wide range of facilities (some of which may be very advanced), they can usually handle large data sets, they usually have good documentation (many have on-line help facilities) and their high level natural language commands make them comparatively easy to use.

Computer statistical packages will be discussed below under three headings: (a) spreadsheets; (b) specific purpose statistical analysis packages; and (c) general purpose statistical analysis packages.

Spreadsheets

Spreadsheets are currently one of the most underutilized of the computer facilities available to geographers. They are a type of generic software which can be adapted to any type of statistical analysis which requires repeated calculations to be performed on tables of data (Sawicki 1985, Soper and Lee 1987). Their most frequent use to date has been in the area of numerical forecast modelling, although many other uses can be foreseen. Sipe and Hopkins (1984) discuss fifteen spreadsheet models for economic base analysis, small area population projections and budget monitoring; Landis (1985) describes their use for shiftshare analysis of employment trends in Providence County (Rhode Island, USA); Levine (1985) discusses their use in the analysis of Los Angeles County population data; and Lee and Soper (1987) advocate their use in teaching statistics to geographers. The great success of spreadsheets is demonstrated by the fact that the Lotus 1–2–3 spreadsheet package is currently the world's best selling microcomputer application package.

A spreadsheet can be thought of as a large collection of boxes (often called cells or slots) which can be linked together to interact across and down an electronic worksheet. Spreadsheets normally contain several hundred slots. Individual slots are referenced using a label, usually consisting of a letter to identify the column and a number to identify the row (Fig. 4.3). Since most computer screens are restricted to 80 columns × 24 rows of text characters, it is not possible to display the whole of a large sheet at any one time. It is customary, therefore, to display only a portion of the sheet (perhaps the first eight characters of slots A1 to I19). The position of this 'window' on the sheet can normally be moved by using the cursor keys of the computer.

The links between boxes are established when a sheet model (sometimes simply called a sheet) is set up. Links are entered as formulae, for example, the formula in slot G3 in Fig. 4.3 is B3 + C3 + D3 + E3. This means add together the contents of slots B3, C3, D3 and E3 and place the result in G3. Once data are entered into a sheet, calculations are carried out automatically and quickly. Instead of a formula or a data item, a slot may be filled with text which may be used to provide column headings and other labels. The input and editing of slots is achieved by identifying a slot using the screen cursor and then entering or modifying the slot contents. A press of the RETURN key usually terminates input or editing and moves the cursor on to the next slot. When the RETURN key is pressed, the computer updates the whole of the sheet, including any changes which occur from editing. Once a spreadsheet has been created either the whole or part of a sheet can be saved on disk for future use.

	A	B	C	D	E	F	G	H	I
1		1981	1982	1983	1984		TOTAL	MEAN	
2	MONTH								
3	JAN	3	3	2	3		11	2.75	
4	FEB	5	4	3	4		16	4	
5	MAR	7	6	5	6	=	24	6	
6	APR	10	9	8	6	=	33	8.25	
7	MAY	11	12	11	10	=	44	11	
8	JUN	11	11	13	12	=	47	11.75	
9	JUL	13	12	15	11	=	51	12.75	
10	AUG	13	14	17	14	=	58	14.5	
11	SEP	10	11	15	12	=	48	12	
12	OCT	8	8	10	9	=	35	8.75	
13	NOV	6	7	5	8	=	26	6.5	
14	DEC	4	5	3	4	=	16	4	
15									
16		-------	-------	-------	-------		-------		
17	TOTAL	101	102	107	99				
18	MEAN	8.41667	8.5	8.91667	8.25			8.52083	
19									

Fig. 4.3. A simple spreadsheet model for calculating average annual and monthly temperature values.

In addition to the features described above, spreadsheets have a range of other more specialist facilities and features. Most spreadsheets have facilities to assist sheet set up and data entry. These allow users to replicate slot contents, to provide windows so that important parts of large sheets can be displayed together and to protect parts of sheets, such as complex formulae, and so avoid accidental deletion. Some spreadsheets allow users to control access to data files on disk (that is, read and write data from and to disks while the sheet is calculating). This may be important when dealing with large data files. The best spreadsheets can also be interfaced to other types of software, such as data base management system packages, word processing packages and graphics packages.

Creating a spreadsheet to analyse examination data

The analysis of examination marks is a task which occupies a considerable amount of lecturers' time each year. Figure 4.4 is an example of a spreadsheet model, for simple statistical analysis of examination data, that may be able to assist in this regard. This type of analysis is common to several other areas of geography and so the principles may be more widely applied. The example discussed here was prepared using View-Sheet on an Acorn BBC microcomputer, but similar results could have been obtained using any one of a number of other combinations of

(a)

NAME	PAP1	PAP2	PAP3	PAP4	CSWK	MEAN	HISTOGRAM
----------	----	----	----	----	----	----	----¦----¦----¦----¦
BROAD,C	53	48	50	42	66	54.2	**********
BOTHAM,I	54	49	54	53	60	54.7	**********
DOWNTON,P	53	55	56	53	65	57.6	***********
GATTING,M	49	57	54	52	59	54.9	**********
GOWER,D	60	60	58	55	57	58.2	***********
HAYNES,D	61	56	63	54	65	60.9	************
HICK,G	37	54	45	42	60	49.4	*********
IMRAN	49	50	53	48	55	51.7	**********
LAMB,A	55	54	53	50	24	44.6	********
MOXOM,M	65	69	74	71	70	69.7	**************
PARKER,P	40	45	43	46	0	30.2	******
ROEBUCK,P	29	34	40	39	80	48.5	*********
SIMMONS,J	73	76	67	74	85	76.1	***************
SMITH,D	54	58	55	49	53	54.2	**********
----------	----	----	----	----	----	----	----¦----¦----¦----¦
MEAN	52.3	54.6	54.6	52.0	57.1	54.6	**********

(b)

NAME	PAP1	PAP2	PAP3	PAP4	CSWK	MEAN	HISTOGRAM
----------	----	----	----	----	----	----	----¦----¦----¦----¦
BROAD,C	53	48	50	42	66	54.2	**********
BOTHAM,I	54	49	54	53	60	54.7	**********
DOWNTON,P	53	55	56	53	65	57.6	***********
GATTING,M	49	57	54	52	59	54.9	**********
GOWER,D	60	60	58	55	57	58.2	***********
HAYNES,D	61	56	63	54	65	60.9	************
HICK,G	41	54	45	42	60	50.2	**********
IMRAN	49	50	53	48	55	51.7	**********
LAMB,A	55	54	53	50	24	44.6	********
MOXOM,M	65	69	74	71	70	69.7	*************
PARKER,P	40	45	43	46	51	45.5	*********
ROEBUCK,P	29	34	40	39	80	48.5	*********
SIMMONS,J	73	76	67	74	85	76.1	***************
SMITH,D	54	58	55	49	53	54.2	**********
----------	----	----	----	----	----	----	----¦----¦----¦----¦
MEAN	52.6	54.6	54.6	52.0	60.7	55.8	**********

Fig. 4.4. A spreadsheet model for calculating examination statistics. See text for explanation of differences in (a) and (b).

hardware and software. The sheet comprises a series of windows containing:

1. the list of names;
2. the marks for the papers;
3. the mean mark for each candidate;

4. the bar chart of mean marks;
5. the text below the name window;
6. the mean papers marks;
7. the class mean; and
8. the class bar.

The windows are defined by their top left and bottom right position in the sheet. The column width of the windows has also been set. Window (1), for example, extends from slot A1 to A12 and has a column width of 10 characters. The statistics are displayed by writing formulae in slots, for example, G3 contains the formula (B3/5) + (C3/5) + (D3/5) + (E3/10) + (F3/10×3). This has the effect of calculating the weighted mean for each of the percentage marks achieved by Broad. The papers are weighted as follows: PAP1 20 per cent, PAP2 20 per cent, PAP3 20 per cent, PAP4 10 per cent, CSWK (coursework) 30 per cent. The bar charts are drawn by requesting the option in the window description.

The real value of spreadsheets can be demonstrated when alterations or additions need to be made to the data. Two alterations to the examination data might be: first, the external examiners may decide to increase Hick's mark of 37 per cent on PAP1 to 41 per cent and, second, to allow the mark for Parker's coursework (which was submitted late) to be counted. The effect of these two changes can be seen by comparing Figs 4.4 (a) and (b). Note the eight changes in the summary statistics which arise from two changes in the data. The whole process of data editing and recalculation took only a few seconds.

Specific purpose statistical analysis packages

Specific purpose statistical analysis packages are specific in the sense that their sphere of use is restricted to the analysis of one type of problem. There are many such packages available to geographers for a great variety of tasks too numerous to mention here. Examples include the vegetation analysis packages TWINSPAN and DECORANA (Hill 1979a, b), widely used for classifying vegetation communities using multivariate statistics. The GLIM (Generalized Linear Interactive Modelling) package already described in Chapter 1 is a further example. GLIM can be used for various types of linear modelling using different types of data (O'Brien and Wrigley 1980). Many specific purpose packages have also been developed for survey analysis and some of these will now be discussed.

Survey analysis packages have many of the characteristics of the statistical analysis software already described. In particular they are closely allied to spreadsheets. Survey analysis packages are worthy of consideration here because of their utility in the analysis of questionnaire-type data and other social science surveys (for example telephone interviews). A number of survey analysis packages are currently available for computers and brief details of these are given in Weatherill and Curram (1984), Gilchrist (1985) and Cable and Rowe (1987). The use of the SNAP survey analysis package is described in more detail below. In general, survey analysis packages have facilities for data entry, listing and editing, as well as analyses such as cross tabulations, cell counts, descriptive statistics (mean, median, standard deviation etc.), inferential statistics (chi-square, regression, ANOVA etc.) and graphics (mainly histograms and bar charts). Most survey analysis packages can be run in either batch or interactive mode.

SNAP

SNAP is a survey analysis package developed by Mercator Computer Systems. It was launched in 1980 and is now available for a range of machines. The program is menu driven, relatively easy to use and comes with reasonable documentation. It is well supported by the authors and is frequently updated.

Data entry to SNAP is either from the keyboard, from hand-held portable terminals or via telephone links to other computers (see Chapter 9). Data can be entered in an uncoded format direct from questionnaires or in a precoded format. Both numerical and alphanumerical input is allowed. A wide variety of analyses are possible including cell counts, cross tabulations, descriptive statistics and bar graphs (Fig. 4.5). A number of filters (for example, subset females from the data) and weights (for example adjust records to reflect composition of a sample) can be applied. These can be undertaken in either batch or interactive mode. Output may be to the screen, disk or dot matrix printer. Unfortunately SNAP provides only limited functions for statistical analysis. It does not, for example, have chi-square, regression and factor analysis and the graphics are also limited. Wilding (1985) provides a more extended review of SNAP.

General purpose statistical analysis packages

General purpose statistical packages provide geographers with many of the basic statistical techniques which are required for the vast majority of geographical research. Of the packages which are available, SPSS (Statistical Package for the Social Sciences) was the most popular in the 1970s and is still in common use. It is especially useful for the analysis of social survey data. BMDP and Genstat were also commonly used in the 1970s. However, in the last few years the Minitab package has gained increasing favour with geographers, primarily because of its ease of use,

SNAP

DEMONSTRATION SURVEY

DEMO REPORT

ROW : REGION (1)
COLUMN : AGE/SEX/SEG (39)
FILTER : MALE/FEMALE (3) -x1x
 % BY : COLUMN

| AGE/SEX/SEG | / | / | / | / | / | / | / | / |
REGION	/ Totals/	MALE	/FEMALE/	AGE / 0-26/	AGE / 26-99/	GROUB/ AB /	GROUP/ C1/C2/	GROUP/ DE /
Totals /	265	265	0	60	205	25	220	20
LONDON /	21.1%	21.1%	0.0%	0.0%	27.3%	20.0%	14.1%	100.0%
/	56	56	0	0	56	5	31	20
SOUTH EAST /	18.5%	18.5%	0.0%	0.0%	23.9%	80.0%	13.2%	0.0%
/	49	49	0	0	49	20	29	0
SOUTH /	21.9%	21.9%	0.0%	66.7%	8.8%	0.0%	26.4%	0.0%
/	58	58	0	40	18	0	58	0
SOUTH WEST /	17.4%	17.4%	0.0%	33.3%	12.7%	0.0%	20.9%	0.0%
/	46	46	0	20	26	0	46	0
MIDLANDS /	2.3%	2.3%	0.0%	0.0%	2.9%	0.0%	2.7%	0.0%
/	6	6	0	0	6	0	6	0
NORTH EAST /	0.4%	0.4%	0.0%	0.0%	0.5%	0.0%	0.5%	0.0%
/	1	1	0	0	1	0	1	0
NORTH WEST /	17.0%	17.0%	0.0%	0.0%	22.0%	0.0%	20.5%	0.0%
/	45	45	0	0	45	0	45	0
WALES /	1.5%	1.5%	0.0%	0.0%	2.0%	0.0%	1.8%	0.0%
/	4	4	0	0	4	0	4	0
SCOTLAND /	0.0%	0.0%	0.0%	0.0%	0.0%	0.0%	0.0%	0.0%
/	0	0	0	0	0	0	0	0

Fig. 4.5. Cross tabulation between socio-economic group (SEG) and region
produced using SNAP.
Source: provided by Mercator Computer Systems, Bristol.

and it is now the most frequently used of the general purpose statistical analysis packages. An example of the application of Minitab for geographical research will now be presented.

The analysis of geographical data using Minitab

The Minitab general purpose statistical analysis package was originally designed in 1972, for use by students in introductory statistics courses at Pennsylvania State University (Ryan, Joiner and Ryan 1985). It is an interactive package that has facilities for descriptive and inferential statistics, as well as graphics and exploratory data analysis. There are commands to read, edit and print data, and to do arithmetic and transformations. Minitab has been implemented on a wide range of computers. It is an extremely easy to use and well documented computer package, which stands as an example to all software developers.

The central Minitab concept is that of the worksheet (rather like a spreadsheet or table). In a Minitab worksheet (Fig. 4.6) each row represents a case (for example a rainfall station or a respondent to a questionnaire) and each column represents a variable (for example the rainfall or the age of a respondent). An individual worksheet can usually contain up to several hundred rows and columns. Statistical operations can be performed on the data in a worksheet using commands in the form of keywords. For example, PRINT C1–C2 will print the contents of columns 1 and 2 from the worksheet, and DESCRIBE C1–C2 will produce descriptive statistics of the two columns (Fig. 4.6).

The computer output shown in Figs 4.6 and 4.7 demonstrates how Minitab can be used to carry out a typical geographical investigation (only the basic points relevant to interpreting Minitab output will be considered here – a statistical text such as Silk 1979 should be consulted for further details of the methods employed). The objective of this investigation is to try to describe, explain and predict spatial variations in rainfall in Northumberland, UK.

It is suspected that altitude may play a part in determining the level of rainfall. Thus data have been collected about the mean annual rainfall, from twenty-one stations at different altitudes, and input into Minitab (in Fig. 4.6 the data are shown being read in from a file called 'NERAIN. DAT' after having been input into Minitab and then saved). A scatterplot showing the relationship between the data (Fig. 4.7) indicates that there is a fairly strong relationship between the level of rainfall and altitude. It also shows the linearity of the trend and that the conditional distributions have approximately equal variances (two of the assumptions of regression analysis). Regression analysis can now be used to examine the relationship between rainfall and altitude more objectively.

```
OK, MINITAB

MTB > READ 'NERAIN.DAT' C1-C2
     21 ROWS READ
  ROW     C1      C2

    1     640      46
    2     620      11
    3     670      53
    4     810     183
    .   .   .

MTB > NAME C1 'RAINFALL' C2 'ALTITUDE'
MTB > PRINT C1-C2
  ROW   RAINFALL   ALTITUDE

    1        640         46
    2        620         11
    3        670         53
    4        810        183
    5        785        104
    6        730        111
    7        730         76
    8       1014        373
    9       1055        375
   10        895        267
   11        810        213
   12        747         99
   13        860        177
   14       1050        266
   15        810        198
   16       1199        290
   17        945        122
   18        972        327
   19        856        267
   20        771        162
   21        854        222

MTB > DESCRIBE C1-C2

              N      MEAN    MEDIAN    TRMEAN    STDEV
RAINFALL     21     848.7     810.0     842.3    148.5
ALTITUDE     21     187.7     183.0     187.2    106.7

           SEMEAN       MIN       MAX        Q1       Q3
RAINFALL     32.4     620.0    1199.0     738.5    958.5
ALTITUDE     23.3      11.0     375.0     101.5    267.0
```

Fig. 4.6. Output from the Minitab general purpose statistical analysis package. The descriptive statistics shown are in order: the number, mean, median, trimmed mean (mean of the middle 90 per cent of the data range), standard deviation, standard error of the mean, minimum data value, maximum data value, lower quartile and upper quartile.

Regression is performed using the command REGRESSION C1 1 C2, which means that the dependent variable (RAINFALL) is in column 1 and that there is 1 explanatory (Predictor) variable (ALTITUDE) in column 2. Minitab first prints the regression equation, followed by the coefficient table. The first line relates to the coefficient

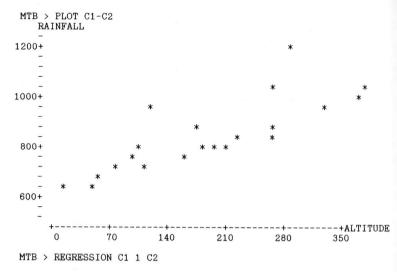

```
MTB > PLOT C1-C2
    RAINFALL
     -
1200+                                        *
     -
     -
     -                              *              *
1000+                                              *
     -              *                        *
     -
     -                    *              *
     -                         *    *
 800+         *          * * *
     -        *        *
     -     *      *
     -        *
     -  *     *
 600+
     -
     +---------+---------+---------+---------+---------+ALTITUDE
     0        70       140       210       280       350
```

MTB > REGRESSION C1 1 C2

```
The regression equation is
RAINFALL = 628 + 1.18 ALTITUDE

Predictor       Coef         Stdev       t-ratio
Constant        628.12       36.69        17.12
ALTITUDE        1.1752       0.1709        6.88

s = 81.55       R-sq = 71.3%     R-sq(adj) = 69.8%
```

MTB > STOP

Fig. 4.7. Output from Minitab (continued). See text for explanation.

α (Constant) and the second to β (ALTITUDE). The standard deviation (Stdev) of a coefficient (Coef) is a measure of its confidence interval. The ratio (*t*-ratio) between a coefficient and its standard deviation can be used to estimate the level of confidence we can place in a coefficient value. As a rule of thumb, a *t*-ratio greater than 2.00 indicates that we can place a high degree of confidence in the value of a coefficient. The s value is the standard deviation of the dependent variable (rainfall), a measure of the confidence limit we can place around any prediction of rainfall using altitude. The R-sq is a measure of how well the regression equation fits the data. It is a measure of the statistical explanation of rainfall which may be attributed to altitude. R-sq(adj) shows the R-sq value adjusted for the size of the data set. Further details about the interpretation of Minitab output are given in Ryan, Joiner and Ryan (1985).

The fact that the R-sq value at 71.3 per cent is relatively high, the *t*-ratio values at 17.12 and 6.88 are relatively high and the s value at 81.55 is relatively small in comparison to the values for rainfall (range

620–1,199 mm), means that we can be confident in using altitude values to describe, explain and predict spatial variations in rainfall in North-umberland. (Statistical purists may wish to note that the data are slightly heteroscedastic. This can be removed by log transformation of the dependent variable (RAINFALL) which increases the R-sq valued to 74.6 per cent.) Thus in future research, altitude values, which can be easily and cheaply obtained from topographic maps, can be used as a surrogate for rainfall, which is in comparison expensive to collect using rain gauges.

CONCLUSION

Statistical analysis has traditionally been a very strong element of quantitative geography. However, in recent years the trend has been away from pure research concentrating on developing and refining methods toward the incorporation of existing methods in other types of geographical computer applications. Computer cartography, remote sensing, simulation and geographical information systems are all examples of geographical computer applications which make great use of statistical techniques.

The use of statistics in geography has always been closely linked with computers and some of the best computer software has been developed for statistical analysis. Many would agree that the Minitab statistical analysis package is one of the best available to geographers and other scientists and social scientists. Continuing software development will ensure that this remains so in the immediate future.

This chapter has reviewed the main types of statistical analysis software currently available and has presented two examples which demonstrate how statistical analysis can be of assistance to the geographer. This basic appreciation of statistical analysis provides the foundation for an understanding of the principles and applications of computer cartography, remote sensing and simulation, the subjects to be covered in the next chapters.

FURTHER READING

Cohen, L., Holliday, M. (1982) *Statistics for Social Sciences*. Harper and Row. (A good text on statistics with a comprehensive collection of BASIC computer programs.)

Ebdon, D. (1985) *Statistics in Geography*. Basil Blackwell. (A clear and concise if rather limited statistics text with BASIC computer programs and several good worked examples.)

Gilchrist, R. (1985) Statistical packages for the IBM PC. In **Barnetson, P.** (ed) *The Research and Academic Users Guide to the IBM Personal Computer*. IBM UK Ltd, pp 111–29. (Although aimed at the IBM PC it contains a useful review of statistical analysis software for microcomputers.)

Ryan, B. F., Joiner, B. L., Ryan, T. A. (1985) *Minitab Handbook* 2nd edn. Duxbury Press, Boston. (As well as giving full details about Minitab, this is a very useful elementary statistics work book.)

Unwin, D. J., Dawson, J. A. (1985) *Computer Programming for Geographers*. Longman. (Much of general interest; see especially Chapter 9 on files and packages.)

5

Computer cartography

This chapter is concerned with the creation of cartographic and graphic images using computers. The process of creating cartographic and graphic images has been variously referred to as computer-aided, computer-assisted and automated cartography, but because of the confusion which arises from the use and misuse of this terminology, the simple term computer cartography is preferred here. Strictly speaking, computer cartography is only concerned with maps and map-like products, but since the same basic principles also apply to the creation of graphic images, such as statistical graphs, they are also included here.

Despite the relatively recent nature of computer cartography it is already a large-scale activity. It is worth reflecting on the fact that, because of the widespread use of computer cartography by the media (for example newspaper and television companies), more cartographic and graphic images are now produced by computer than by traditional manual means. The massive recent expansion in the importance of computers in cartography has also been highlighted by Joel Morrison (1986), President of the International Cartographic Association, who suggests that computer cartography should now be regarded as mainstream cartography and that older technologies, such as scribing, printing and historical map analysis are more marginal.

The various methods which people use to develop, analyse and communicate ideas have been described as the four 'acies' or 'aces' for short (Balchin and Coleman 1965). The four aces are literacy (written language), numeracy (mathematical symbols), oracy (spoken word – sometimes called articulacy) and graphicacy (cartographic and graphic images). Balchin (1985) has further described graphicacy as the communication of spatial information that cannot be conveyed adequately by written or numerical means. Many people argue that the concern for graphicacy is one of the things which sets geography apart from other sciences and social sciences.

The first part of this chapter discusses the basic principles of computer cartography. The next section on computer cartography hardware and software, includes discussion of a range of hardware devices and software packages. Following this the advantages and disadvantages of computer cartography are outlined. There is then brief coverage of two of the major applications of computer cartography, namely, topographic and thematic mapping, and statistical graphics (image analysis and geographical information systems are covered in later chapters). The final section is a case study of mapping population census statistics using the GIMMS computer cartography package.

THE PRINCIPLES OF COMPUTER CARTOGRAPHY

Vector and raster graphics

The two major systems for creating cartographic images using a computer are called *vector* and *raster* graphics (Fig. 5.1). In vector graphics, images are built up from points, lines and areas, each defined by pairs of $x-y$ co-ordinates. In raster graphics images are built up from cells called *pixels* (picture elements). There are differences in the way vector and raster data are collected, stored, manipulated and displayed. Both systems have advantages and disadvantages depending on the way in which the cartographic data are to be used. Vector graphics are usually employed where it is necessary to integrate manual and computer cartography techniques and where topological structure and annotations are required. The vector system also has the advantage that it is a more efficient data storage structure than the raster system. In the vector system, only the co-ordinates which actually describe the features in a cartographic image need to be coded. In the raster scheme every pixel in the image must be coded as either full or empty (compare the amount of information which needs to be stored for the image displayed in Fig. 5.1), although there have been a number of recent advances which have substantially improved the efficiency of raster coding (Burrough 1986). The difference in the amount of storage capacity is not significant for small images, but for large images it may be a major consideration. Raster graphics are usually used where it is necessary to integrate topographic and thematic map data with remote sensing data (which is

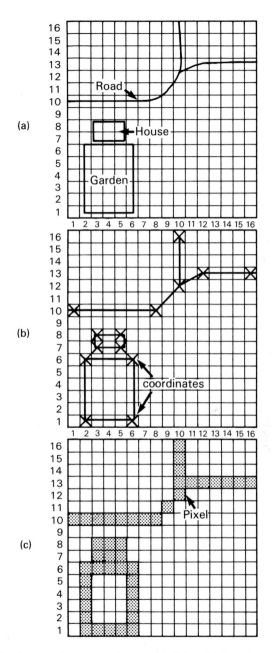

Fig. 5.1. Systems for representing graphical data: (a) in analogue form on a map; (b) in digital vector form on a computer; (c) in digital raster form on a computer.

stored in raster format – see Chapter 6). Image comparison, area filling and light pen operations (see below) are also best achieved using raster graphics. Microcomputers and many computer screens based on television technology use the raster system. The two systems are best viewed as complementary rather than competitive. This is especially so since vector-to-raster and raster-to-vector conversion routines are now available, making it feasible to carry out different operations using the same data in either raster or vector format. The advances in geographical information systems reported in Chapter 10 also mean that it is now possible to integrate both vector and raster data more easily.

Resolution

An important measure of the cartographic capabilities of a computer system is the *resolution*. This is normally expressed as the number of discrete units (pixels or co-ordinate values) along the x and y axes which can be used to store or display an image. It is customary to classify computer cartography systems as low, medium and high resolution, but there does not seem to be general agreement about the precise definition of these terms. This is hardly surprising since the resolution of computer systems is constantly increasing as a result of advances in technology. At present it is convenient to think of low resolution as less than 100,000 pixels or co-ordinate values, medium resolution as 100,000–250,000 and high resolution as greater than 250,000. It is interesting to note that Carter (1980), in his discussion of cartographic resolution, suggests $1,024 \times 780$ (798,720) pixels as a minimum acceptable to many viewers. This is also the resolution of the Tektronix 4010, often regarded as the standard computer cartography terminal. For comparison, the IBM PS/2 microcomputer has a resolution of 640 $\times$ 480 (307,200) pixels and for image analysis systems 512×512 (262,144) pixels is regarded as the preferred minimum.

COMPUTER CARTOGRAPHY HARDWARE

The purpose of this section is to present a brief introduction to the main types of specialist hardware devices which are used for computer cartography, in addition to the basic components of a computer system (keyboard, processor, disk drive and screen – see Chapter 11). The basic additional devices for cartographic input are the digitizer, joystick, mouse and light pen. Further, less common cartographic input devices, such as the trackball and paddle, are briefly described in the Glossary.

The basic additional devices for hard copy (permanent) cartographic output are the plotter, camera and printer (see Chapter 8 for a discussion of printers).

Digitizer

A digitizer is a device for capturing conventional analogue topographic and thematic maps in a format suitable for storage and manipulation in a computer. A digitizer may be used to capture data in either vector or raster format.

Vector format data may be captured by either electromechanical or semiautomatic line follower digitizers. An *electromechanical digitizer* is essentially a tablet of electronic 'graph paper' (Fig. 5.2). Attached to the tablet is a pen or tracking cross which can be moved around the tablet and can detect the signal at any intersection of a grid of wires. This analogue signal is coded by the computer into a digital x–y co-ordinate, measured absolutely from a user-defined origin. Electromechanical digitizers may work in either point or line/stream mode. In point mode x–y co-ordinates are recorded only when an operator gives a signal, such as the push of a button. In line/stream mode the digitizer is set up to record co-ordinates at fixed time or distance intervals. Electromechanical digitizers are normally available up to about A0 (0.841 m ×

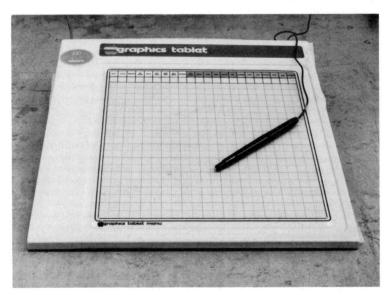

Fig. 5.2. A small digitizer.

1.189 m) size and their resolution (the distance between the wires in the grid) is normally in the range 0.01–0.1 mm. More recently, smaller low cost variable resistor electromechanical digitizers have become available. These consist of a cross on the end of an articulated arm, the position of which can be determined by a variable resistor. Although cheaper, these digitizers are less accurate, more fragile and less pleasant to use. A *semiautomatic line follower digitizer* uses a photosensor or fine laser beam to follow the lines on maps and records co-ordinates at fixed time or distance intervals. Operator intervention is required to start the process at the begining of a line and to decide upon which course to follow at the intersection of lines.

The two main types of raster format digitizers are scanning densitometers and electronic video systems. In a *scanning densitometer* the map is fixed to a drum which rotates beneath a photosensitive optic. The map is captured in the form of an image which consists of a matrix of pixels. The optic records the intensity of light for each of the pixels on a line as the map rotates. At the end of each scan line the scanner optics advance one scan width down the map. Scanning densitometers have a typical resolution of up to 0.012 mm and can collect data at the rate of 30,000 pixels s^{-1}. An *electronic videodigitizer* essentially uses a video camera to grab an image usually as a collection of up to 512×512 pixels each containing a level of light intensity. At present electronic video systems are cheaper than scanning densitometers, but they cannot attain the same resolution.

Each type of digitizer has advantages and disadvantages depending upon the type of application for which it is to be used. Generally speaking, electromechanical digitizers are the most common because of their relatively low capital cost, ease of use, high resolution and because they produce data in vector format which is still the most common format used in computer cartography. In situations where very large amounts of digitizing must be undertaken, for example in national mapping programs, then semiautomatic vector and raster format digitizers are often used since they are much faster and in such circumstances are cheaper. If necessary, data output from a raster format digitizer can be fairly easily converted into vector format.

Joystick

A joystick is a lever which can be moved in two dimensions (Fig. 11.10). Two potentiometers (normally at right angles) sense this movement and convert it to *x–y* co-ordinates. Joysticks are often used to position a screen cursor, to input commands from a screen menu or locate the corners of a screen window. The speed of cursor movement can also be controlled by a joystick.

Light pen

A light pen is a small hand-held, pen-like device with a light-sensitive tip. By pressing the tip against a computer screen a light pen can be used to interact with images and programs. Light pens may be used to draw/remove lines or shading on a cartographic image, or input data to a program. The main problems with light pens are their high cost, slow speed, poor accuracy and the fact that holding a pen to a screen can be physically tiring. As ever, these drawbacks are cost-related since the more expensive pens are faster and more accurate.

Plotter

Plotters are the most widely used devices for producing high quality, hard copy computer output. Several types of plotter exist and they all work in different ways. *Drum plotters* were the first type to be developed and they are still the most commonly used by large computer installations. In a drum plotter the paper, which is of fixed width and great length, rolls over a drum that acts as the plotting surface. The image is created by one or more pens which move in the y direction, as the paper moves in the x direction across the drum. Drum plotters work relatively quickly at typical speeds of $40 \, \text{cm s}^{-1}$, they can produce large-size, multicolour, high quality output and are generally fairly reliable.

Flat bed plotters (Fig. 5.3) typically have A3 ($0.297 \, \text{m} \times 0.420 \, \text{m}$) or A4 ($0.210 \times 0.297 \, \text{m}$) sized plotting areas, facilities for multicolour plots and operate at speeds of around $25 \, \text{cm s}^{-1}$. Flat bed plotters are distinguished from other plotters in two major respects. First, the paper is fixed to a flat surface and, second, the pen(s) is supported on a gantry which moves in both an x and y direction. They have a number of useful features such as:

1. they can accept any type of drawing material, including paper and acetate film;
2. they can use liquid ink, ball point, fibre-tip and spirit based pens;
3. some can also be used as digitizers.

Comparatively slow, A4, single pen plotters can be purchased relatively cheaply.

Electrostatic plotters (sometimes called laser printers) operate in a similar fashion to photocopiers. Electrodes deposit charges on chemically treated paper in the form of an image. The paper then passes through a toner bath which adheres black toner to the charged areas. Electrostatic plotters are very fast (after processing the data into a suitable format, plotting takes only a few seconds), they are commonly of medium resolution, but they are capable of producing only monochrome images.

Fig. 5.3. An A3 size flat bed plotter.

Camera

Cameras are increasingly being used as hard copy output devices for computers. At its simplest this may involve using a 35 mm camera to take a picture of a screen. Many of the photographs in this book were taken this way. This technique, however, can suffer from edge effects, screen reflectance and lengthy processing times. The final problem may be circumvented by the use of Polaroid cameras. More acceptable pictures can be obtained by connecting cameras directly to the video or RGB output socket of a computer.

COMPUTER CARTOGRAPHY SOFTWARE

The last twenty-five years have seen the development of a large number of computer cartography programs and packages. This software is of varying levels of sophistication and may be used on a wide range of machines in a great variety of applications. Some of this software is in the form of low level, machine-orientated drawing commands and routines, but the most useful is in the form high level, user-orientated packages.

The genesis of computer cartography, as we know it today, was in the later 1950s and its incorporation into geography was signalled by Tobler's (1959) seminal article on 'Automation and Cartography'. At the time, the major technical problem to overcome was the development of an output method which could produce an easily readable map. This problem was solved by producing cartographic images on standard text screens and line-printers (Maguire 1985a). The computer was used to print out an appropriate symbol in each print position on each row (Fig. 5.4), a type of raster graphics. A number of packages were produced which employed this method, by far the best known of which was SYMAP (Dougenik and Sheehan 1975).

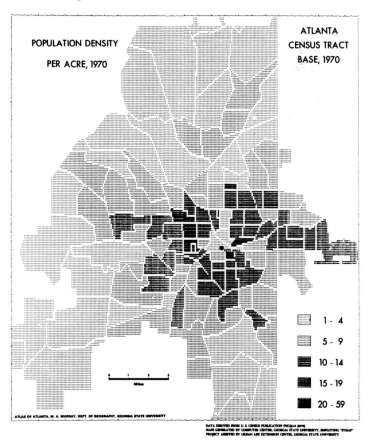

Fig. 5.4. Output from the SYMAP computer cartography package. Note that the lettering and scale bar were not drawn by computer.
Source: Murray 1974.

In the 1970s computer cartography software was developed which used vector graphics and plotters as the principal hardcopy output device. This enabled maps to be drawn that were constructed of points and lines rather than characters. Maps of this type, although more expensive to produce, were of a much higher and more acceptable quality. In this second phase of the development of computer cartography the software was of two basic types, namely, *subroutine libraries* and *specialist packages*. The arguments for and against subroutines and packages have already been rehearsed in the previous chapter, in the context of statistical analysis software, and so they need not be restated here. The best known of the subroutine libraries are DISSPLA, GINO, GHOST and the NAG graphical supplement. The best known of the specialist packages are GIMMS, MAPICS, ODYSSEY, SAS/GRAPH and TELL-A-GRAF (Monmonier 1982; Carter 1984; Jones, T. 1985; Burrough 1986). All these are still in common usage and examples of output from GINO, TELL-A-GRAF and GIMMS are shown in Figs 5.5, 5.6, 5.10 and 5.12 and Plate 1.

In the 1980s microcomputers have been one of the major development areas in the field of computer cartography. The availability of microcomputer-based cartographic workstations (digitizer, microcomputer, software and plotter) reduced the cost of computer cartography by a factor of ten and heralded the third phase of computer cartography. Although much of the early microcomputer cartography was relatively unsophisticated and produced only crude low resolution maps, it was deemed adequate for a number of purposes, such as producing quick draft copies of maps and graphs for use in research and

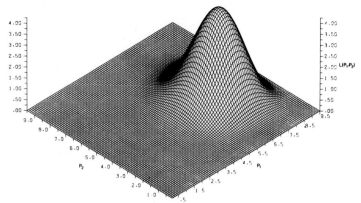

Fig. 5.5. Output from the GINOSURF computer cartography subroutine library.
Source: Vincent and Haworth 1984.

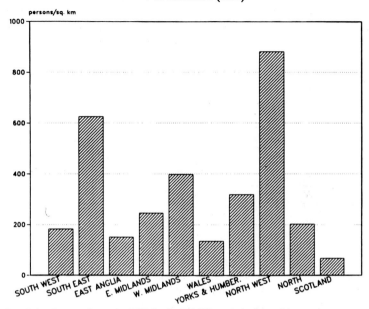

GB STANDARD REGIONS
POPULATION (1981)

Fig. 5.6. Output from the TELL-A-GRAF statistical graphics package.

for teaching purposes. The more recent microcomputer systems, such as ATLAS AMP, MICROMAP, MICROMAPPER, MULTIMAP and PCMAP, are sufficiently advanced to be usable for many geographical applications (Maguire 1985b; Wiggins 1986). Some examples of output from microcomputer cartography systems are shown in Figs 5.7, 5.8 and 5.9. The developments in computer cartography, which resulted from the introduction of microcomputers, profoundly influenced both geography and cartography. In cartography, Taylor (1984) has already proposed that the term 'New Cartography' should be adopted to acknowledge the scale of the changes brought about by the reduction in cost and wider availability of computer cartography. These advances have revolutionized our ability to present information of all kinds in the form of computer-drawn cartographic and graphic images. The role of the computer is certain to increase still further in the future, thereby opening up new and challenging possibilities for the effective communication of information and ideas.

It is impossible to describe all cartographic commands and operations in detail because of the great diversity of hardware, command languages and programming styles used in computer cartography systems. This

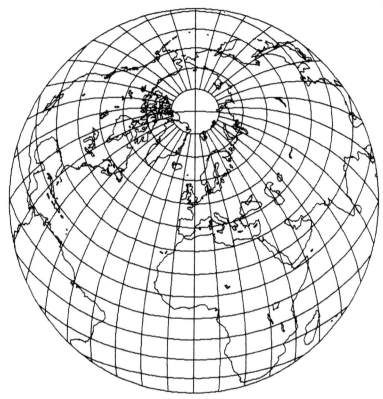

Fig. 5.7. Lambert's Equal Area Azimuthal Projection programmed for the
Atari 800 microcomputer by J. P. Snyder.
Source: provided by J. P. Snyder.

lack of standards in computer cartography is more than an inconvenience for geographers. Hardware manufacturers do not even seem to be
able to agree where the origin (the point at which drawing begins)
should be. For text operations logically the origin should be top left, but
for cartographic operations logically it should be bottom left (as in the
Cartesian co-ordinate system). The lack of standards has arisen for two
major reasons (Brodlie 1985). First, many of the early computer cartography devices, such as plotters, were marketed with their own programs to assist users who wished to draw pictures. Second, around 1970,
a smaller number of 'device-independent' packages were developed (for
example DISSPLA, GHOST and GINO), which subsequently became
de facto standards. Whilst these packages have some advantages, the fact
that there is more than one has led to difficulties in transferring software

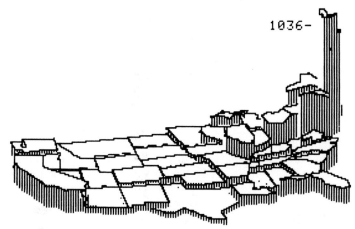

Fig. 5.8. Polyhedron map produced using MICROMAP on an Apple II
microcomputer.
Source: A. R. Jones 1985.

from one type of computer to another. To combat these problems the
Graphical Kernel System (GKS) has been developed as an international standard that is device-independent. This means that a package
written using GKS for use on one system will be easily portable to

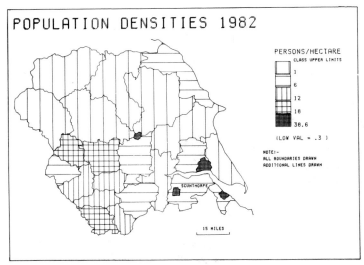

Fig. 5.9. Choropleth map produced using MICROMAPPER on an Apple II
microcomputer.
Source: provided by D. Reeve.

another (Hopgood *et al.* 1983). As far as the user is concerned, the type of computer system should have no effect on the way the software operates.

THE BENEFITS OF COMPUTER CARTOGRAPHY

Rhind (1977) clearly sets out the potential advantages and disadvantages of computer cartography. Although he was concerned solely with computer mapping, most of the discussion is relevant to computer cartography in general. Computer cartography allows existing maps to be produced more cheaply and quickly. The map compilation time is much reduced, storage space is decreased and, most importantly, the cost of updating and redrafting maps held in digital (computer) format is considerably less. Computer cartography facilitates the production of maps which are more closely tailored to user needs. Using a computer cartography system it is relatively simple to produce a map of selected features (for example roads, railways and rivers), for a selected area (the concept of sheet edges does not apply to digital mapping), at any required scale. The advances in computer cartography have led to greater experimentation in the mapping process. This is particularly the case where the maps depict statistical information. Users can easily try out different class boundaries, shading schemes and data transformations for example. Computers allow the creation of new types of cartographic image, not previously possible using traditional means. For example, three-dimensional perspective plots, stereo map pairs and certain map projections all gained in popularity as ways of depicting information because computers made it possible to produce them with ease and speed.

Rhind (1977) does offer a cautionary note by suggesting some of the potential disadvantages which may arise from the introduction of computer cartography. Most computer cartography systems involve a very high initial outlay (often more than was envisaged) and frequently take longer than expected to reach full production. This may be due to hardware or software faults or else due to a lack of familiarity on behalf of the user. This is especially so because many current computer cartography systems are user unfriendly. There may be additional problems about the quality of output from computer cartography systems, since only the most expensive can so far match the drafting standards achieved by professional cartographers. A further danger is that cartographic images will be generated simply because the system is available, rather than because the output is really required. Computer cartography systems are rarely cost effective for producing one-off cartographic images, they are best used in situations where images need to be frequently updated, or where many copies are required with only slight differences. Finally, present computer cartography systems have only limited in-built 'intelligence' and it is quite possible to produce cartographic images, which are geographically incorrect or aesthetically displeasing. The old adage 'garbage in – garbage out' applies equally well to computer cartography as it does to other areas of computer usage.

THE APPLICATIONS OF COMPUTER CARTOGRAPHY

In geography, the principal applications of computer cartography are topographic and thematic mapping, statistical graphics, image analysis and geographical information systems. Thematic and topographic mapping and statistical graphics are discussed below; image analysis and geographical information systems are discussed separately in subsequent chapters.

Topographic and thematic mapping

The objective of *topographic* (also called general) maps is to portray the spatial association of diverse geographical phenomena. They usually show features such as roads, settlements, boundaries, water courses, elevation and coastlines. *Thematic* maps are those which concentrate on the spatial variations of a single phenomenon or the relationship between phenomena (Robinson *et al.* 1984). In practice, as far as computer cartography, is concerned the two types may be considered together since the same basic computer cartography principles are common to both.

Capturing digital map data

The process of capturing and storing digital map data is called digitizing. The simplest form of digitizing involves capturing and storing digital map data in an unstructured format, that is as a collection of co-ordinates in vector format or pixels in raster format. These data can then be replotted on a suitable output device. There are, however, some problems associated with unstructured digital map data and so the majority of current data are captured and stored in structured format (see below). In the case of unstructured digital map data, selective reproduction or manipulation is difficult because it is not possible to identify or gain access to individual map entities, such as rivers, area boundaries and point features. The data storage and processing requirements of unstructured data may be greater than structured data because of the lack of data organization. In the case of unstructured digital thematic map data there may be problems about relating geographical data (point locations, area boundaries etc.) to thematic statistical data. In short, whilst digital map data files in an unstructured format are adequate for reproducing maps, they offer only limited opportunities for data manipulation.

A number of different schemes have been developed for coding maps in vector and raster format in order to create structured digital map data files (Peucker and Chrisman 1975; Burrough 1986, Robinson *et al.* 1984). There is insufficient space to discuss all these here and so two examples of the more commonly used vector and raster techniques will be considered.

The simplest vector coding scheme in which each point, line or polygon is coded independently and completely as a series of *x–y* co-ordinates is called *entity-by-entity* coding. Although very simple to implement, it is grossly inefficient since all common boundaries must be digitized twice (once for each entity). In the more sophisticated *topological* coding scheme, provision is given for coding all common points and boundaries only once. There are several variations of the topological scheme currently in operation. Well known examples are the Dual Independent Map Encoding (DIME) scheme employed by the US Bureau of Census and the 'Harvard' scheme employed by GIMMS and many other packages. In the DIME scheme (Fig. 5.10) maps are

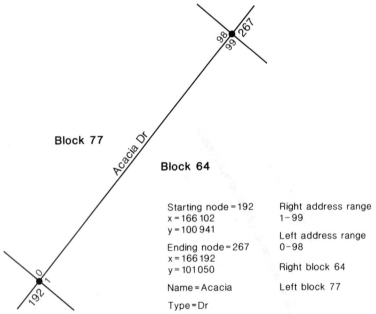

Fig. 5.10. The DIME topological vector coding scheme as used by the US Bureau of Census.

built up from records defining each boundary (street) between adjacent polygons (a block census area). Each polygon boundary is defined by the co-ordinates of the two end-points of the street and the name (census code) of the block on either side.

In the simplest type of raster structure each pixel in an array is referenced by its row and column number (Fig. 5.1) and information is stored about the type of feature. Unfortunately, this is a very inefficient data storage structure. Of the methods which have been developed to reduce the amount of storage space required to hold an image in raster format, the *quadtree* approach has emerged as the most suitable for many applications. In this system an array of pixels describing an image is subdivided hierarchically into four blocks (north-west, north-east, south-west and south-east) until a block wholly occupies an area (Fig. 5.11). The position of a block in a tree and the name of the area it occupies can easily be stored in a computer. The quadtree structure is a very efficient storage structure which is useful for storing and manipulating irregularly shaped areas, but which does not lend itself to the calculation of shape indices and pattern descriptors.

The uses of digital map data

Many countries have national mapping agencies which are responsible for collecting digital topographic map data. For example, in Britain this role is performed by the Ordnance Survey (OS) and in the United States the equivalent is the United States Geological Survey (USGS) Mapping Division. In addition to these, there are many other government, commercial and academic bodies, and individuals involved in the production of digital thematic maps and these may sometimes draw upon government archives.

There are already a number of important data sets available which provide digital map coverage of the world at various scales (Carter 1984; DoE 1987). For example, at the global scale the most well known data sets are the World Data Bank (WDB) I and II files. WDB II was collected from maps at an approximate scale of 1:3,000,000 and contains 6,000,000 co-ordinates describing selected topographic features (Carter 1984). Bickmore (1987) reports on the progress in preparing a World Data Base for environmental sciences by 1990. This topographic data base will contain information such as contours, river networks and coastlines, digitized from maps at a scale of approximately 1:1,000,000. In mid-1986, about 30 per cent of the required data were available. In the United States the whole country is covered at a scale of 1:100,000 and the USGS aims to complete the digitizing of the 1:24,000 (7.5 minute) scale data base by the year 2000 (Starr 1986; Southard 1987). In Britain the whole country is covered at a scale of 1:625,000 and the

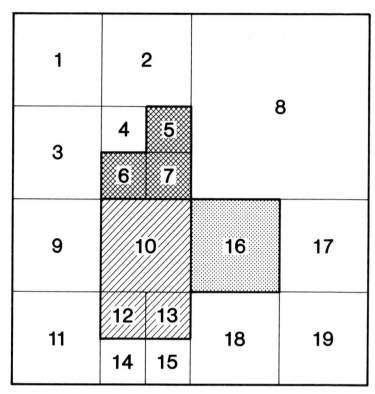

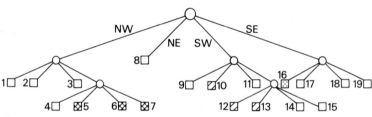

Fig. 5.11. The quadtree raster coding scheme.

OS aims to complete digitizing all the large scale maps (at scales of 1:1250, 1:2500 and 1:10,000 depending on the area) by about the year 2010 (DoE 1987), or perhaps the mid-1990s with the assistance of major users. Much of the work of the national mapping agencies was originally initiated with the simple aim of improving the efficiency of the map production process. It soon became apparent, however, that the data could be used in many other applications such as those discussed below.

Matrices of height data, called Digital Elevation Models (DEMs), are used for many purposes (Burrough 1986; Catlow 1986). The use of a DEM, a slope and aspect algorithm and a model of light reflectance allows the creation of shaded relief maps. In geomorphology, three-dimensional landscape models can easily be created, which facilitate the investigation of soil and water movement on slopes. An appreciation of terrain is also central to many military operations. DEMs are used in the flight simulators in which pilots are trained and in the guidance systems of nuclear missiles.

The utilities (electric, gas, telephone, water) use digital maps of urban areas as a spatial reference to locate pipes, cables and wires. These records can be updated quickly as new installations are added to the networks. They can also be used to co-ordinate inspection and repair operations. In Britain alone, it is estimated that the utilities make around two million openings in the ground each year (DoE 1987). Co-ordination between the different utilities can obviously save considerable amounts of money and greatly reduce the inconvenience to members of the public.

Computer cartography has many applications in the inventory and monitoring of natural resources. In France an inventory of coastal resources has been created from aerial photographs, to help implement a protection policy for the coastal zone (Grelot 1986). Computer cartography is widely used for mineral exploitation. Sheath (1986), for example, describes its use for hydrocarbon exploitation. Many of the areas of the world in which gas and oil occur are extremely sparsely settled or uninhabited and are largely unmapped. There is a need, therefore, for basic topographic maps, so that claims can be registered and located for future reference.

Computers have been used extensively for mapping statistical data, such as the population of cities, the discharge of rivers and vegetation indices derived from the analysis of satellite data. Many atlases have been produced from all over the world using computer cartography techniques (see for example Census Research Unit 1980; Minnesota Department of Energy, Planning and Development 1983; Reeve and Carrick 1983; Hudson, Rhind and Mounsey 1984; Maguire, Brayshay and Chalkley 1987). More detailed coverage of the process of mapping population census data is given later in this chapter in the section on mapping using the GIMMS system.

Statistical graphics

Although statistical (also sometimes called business) graphics have been around for a number of years, the recent more widespread use of computers has been responsible for an upsurge in interest in their use. The term 'statistical graphics' refers to all pictorial methods of presenting statistical information and includes several types of charts, graphs and diagrams. Dickinson (1973) offers four reasons why statistical graphics are used:

1. to arouse interest in the statistics presented;
2. to clarify, simplify or explain statistics;
3. to prove a point referred to in text or speech;
4. to act as a source of information for future analysis.

Statistical graphics are an extremely powerful method of presenting information. This is effectively demonstrated by, for example, the work of Tukey (1977) and Huff (1973). Tukey argues for the increased use of graphical presentations by encouraging wider acceptance of exploratory data analysis, a group of techniques used for data familiarization and exploration. Huff demonstrates the value of graphical presentations in a rather perverse way, by showing how such techniques can be used to mislead the unwary.

The statistical graphics commonly used in geography may be classified as:

1. bar charts and histograms – where the length of each bar is proportional to the value of one data category or class (see Fig. 5.6);
2. pie charts – where a circle (pie) is divided in proportion to the values of two or more variables (see Fig. 5.12);

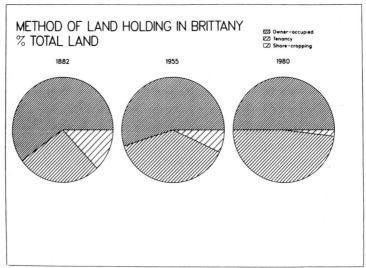

Fig. 5.12. Pie charts produced using the TELL-A-GRAF statistical graphics package.

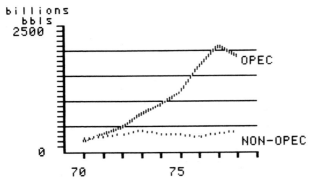

Fig. 5.13. A scatterplot produced using MICROMAP on an Apple II
     microcomputer.
Source: Jones, A. R. 1985.

3. scatterplots and line graphs – where the values of two variables are
   plotted as a symbol (scatterplot) or as a trend line (line graph) (see
   Fig. 5.13).

A number of variations on these themes also exist, for example
population pyramids and pollen diagrams are related to bar charts.

Many statistical analysis and mapping systems offer some statistical
graphics. The facilities offered by Minitab and ViewSheet, for example,
are discussed elsewhere in this book. Other systems which offer facilities for creating statistical graphics include SAS/GRAPH (Green *et al.*
1985), TELL-A-GRAF (Maguire 1986a) and GIMMS (see below).
The advances in microcomputers in recent years have provided much
greater and more immediate access to desk top statistical graphics. A
number of microcomputer packages are reviewed in Bagshaw (1985)
and Moody (1985). Most of these offer a wide range of graphs and
charts, in both monochrome and colour, including some more unusual
plots such as three-dimensional histograms, word charts and curve
fitting using mathematical functions.

Mapping population census data using Gimms

GIMMS (Geographical Information Mapping and Modelling System)
is a high level, command-orientated, computer cartography system that
can be used in both batch and interactive modes (Waugh and McCalden
1983; Carruthers 1985). It was initiated in 1970 by Tom Waugh, a
geographer at the University of Edinburgh, UK. GIMMS has subsequently been updated on several occasions and is currently one of the

most widely used computer cartography packages, with over two hundred sites worldwide. It is available for a wide range of computers from mainframes to microcomputers. GIMMS may be used to create two basic types of cartographic image. First, it may be used to create statistical graphics, such as line graphs, pie charts and scatterplots. Second, it can be used to create maps of points, lines, areas and surfaces. Output from GIMMS has been used to compose maps and graphics in many countries. For example, in England it has been used to map population census data (Maguire *et al.* 1983), in Ireland for agricultural data (Horner, Walsh and Williams 1984), in the European Community for social and economic data (Hudson, Rhind and Mounsey 1984) and in Australia, Canada and the USA amongst others it has been used by government departments for mapping statistical information. Examples of GIMMS output are shown in Plate 1 and in Fig. 5.14.

GIMMS is organized as a series of modules each of which covers

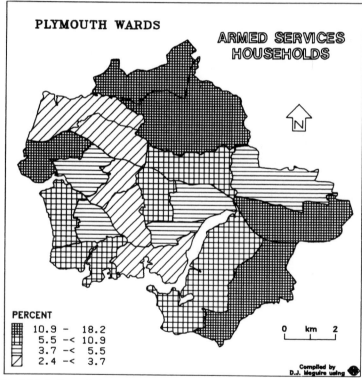

Fig. 5.14. A choropleth map produced using the GIMMS computer cartography package (see also Fig. 5.15).

```
< GIMMS file to draw Armed Services Households map
  Input files PLYMWARDS.POLY,11
              PLYMCENSUS.DAT,12 >

< select plotter as output device >
*PLOTPARM PLOTTER

< enter plotting module >
*PLOTPROG

< define the map size in cm >
*NEWMAP MAPSIZE=14.0 14.0

< read in files >
*GIMMSFILE FILE = 11
*DATAFILE FILE = 12

< define shading >
*SYMBOLISM AREA SPACING=0.25 ANGLE=45   /
                SPACING=0.20 ANGLE=0    /
                SPACING=0.16 ANGLE=0    /
                SPACING=0.10 ANGLE=0  & SPACING=0.10 ANGLE=90 /

< define classes >
*LEVELS = 4
*INTERVALS VARIABLE = ARMED USER = 2.4 3.7 5.5 10.9 18.2

< draw legend >
*LEGEND POSITION = 0.3 0.3 SIZE = 0.4
*TEXT KEY = TOP TEXT = 'PERCENT' ALPHABET = 15

< draw scale bar >
*DRAW 11.0 1.6 11.0 1.4 13.0 1.4 13.0 1.6
*DRAW 12.0 1.4 12.0 1.6
*TEXT POSITION = 10.9 1.7 SIZE = 0.2 TEXT = '0'   ALPHABET = 15
*TEXT POSITION = 11.8 1.7 SIZE = 0.2 TEXT = 'km'  ALPHABET = 15
*TEXT POSITION = 12.9 1.7 SIZE = 0.2 TEXT = '2'   ALPHABET = 15

< draw north point >
*NORTHPT POSITION = 11.5 9.2 SIZE = 1.0 NDRAW

< add titles >
*TEXT POSITION = 1.1 13.0 SIZE = 0.3
TEXT = 'PLYMOUTH WARDS' ALPHABET = 15
*TEXT POSITION = 10.8 12.5 SIZE = 0.4
TEXT = 'ARMED SERVICES
HOUSEHOLDS' ALPHABET=61 CENTRE ONX
*TEXT POSITION = 12.3 0.3 SIZE = 0.15
TEXT = 'Compiled by
D.J. Maguire using' ALPHABET=15 CENTRE ONX

< draw map >
*MAP VARIABLE = ARMED MATYPE = AREA

< leave plotting module >
*END

< leave GIMMS >
*STOP
```

Fig. 5.15. The GIMMS command file used to create the choropleth map
shown in Fig. 5.14.

one aspect of the cartographic image creation process. Thus the GRAPHICS module is used to draw statistical graphics, the FILEIN, COMPILE and PLOTPROG modules are concerned with mapping, and the MANIPULATE and UTILITIES modules deal with data and system organization. In GIMMS, commands comprise a keyword preceded by an asterisk (*) as in *SYMBOLISM, *MAP and *TEXT. Most commands have options; for example, the location of a text string is entered as *TEXT POSITION = 5.0 3.6, where the values 5.0 and 3.6 identify the start location of the text (all measurements in GIMMS are given in cm), measured from the bottom left corner of a map. Comments may be entered between angled brackets (< and >) at almost any stage of a GIMMS command sequence. These are ignored by GIMMS when files are processed.

This section describes the use of GIMMS for producing a map of data from the 1981 Population Census of England and Wales, although many other data sets could have been chosen to illustrate the applications of GIMMS. The aim is to show how to produce a map depicting the percentage number of the armed services households in Plymouth, UK (Fig. 5.14). The background to the Census has already been covered in Chapter 3 and the practice of mapping census data is considered at length in Rhind (1983) and Maguire *et al.* (1984).

The production of an area (choropleth) map using GIMMS requires three files containing locational, non-locational (statistical) and command data. The locational data are $x–y$ co-ordinates (GIMMS works in vector mode), normally collected using a digitizer, which describe a number of areal units. In the example described here the locational data are administrative wards. The statistical data are the actual data to be mapped in the areas. The percentage number of armed services households is a variable extracted from the Census files using the SASPAC (Small Area Statistics PACkage) which was discussed in Chapter 3. Command data are GIMMS commands used to combine the locational and statistical data to produce a map with a title, scale bar, north point and legend.

The discussion here will concentrate on the map production phase, for details of the collection, checking and processing of locational and statistical data files see Waugh and McCalden (1983). The GIMMS command file includes commands to carry out a number of functions (Fig. 5.15). The output device must first be defined, in this case it is to be a plotter (alternatively it may be an interactive graphics terminal). The map-drawing module is then entered and commands are given to define the map size, read in the locational (GIMMSFILE) and statistical (DATAFILE) data files. Next the type of shading, the number of classes and the class boundaries are specified. The SPACING option defines the distance between the shading lines, the ANGLE option defines the orientation of the shading (measured degrees anti-clockwise from east), the & symbol is used to overprint two sets of shading lines, the USER option defines the lower and upper class boundaries for each of the four classes. After this a series of commands are used to add the legend, scale bar, north point and titles. The *TEXT command is used to draw text at a given POSITION, in a given ALPHABET (GIMMS offers a choice of over fifty different alphabets). The *DRAW command is used to draw the scale bar, from the first pair of x–y co-ordinates to the next pair and so on. The CENTRE ONX command centres the text on the x co-ordinate. The final commands are used to leave the PLOT-PROG module and then GIMMS.

To create a map this file must be processed by GIMMS. This involves reading in the commands, which in turn read in the locational and statistical data files and compile the map. The result is a file of low level map-drawing commands that can be used to drive a plotter and create a hardcopy of the map. The real value of the system can be seen when it is realized that a single statistical data file can contain perhaps 50 variables. A map of each one of these can be quickly produced by simply editing the file containing the GIMMS commands. In the case of the Plymouth District example, the only parameters which need to be changed, to produced a comparable map of another variable, are the values for VARIABLE, USER, and the map title identified by TEXT. These are underlined in Fig. 5.15.

CONCLUSION

Without doubt computers have revolutionized the map-drawing process and are now in use both in place of and as a supplement to conventional cartographic techniques in many academic, commercial and government organizations. This chapter has outlined the basic principles, has described the salient characteristics of hardware and software and has discussed some of the applications of computer cartography. It is clear that current advances in computer technology and software development will serve to heighten the dependency of the cartographer on the computer. However, it is also true that cartographic theory must be re-examined and advanced if cartographers are to exploit fully the great potential of the computer.

FURTHER READING

Batty, M. (1987) *Microcomputer Graphics: Art Design and Creative Modelling.* Chapman and Hall. (Some interesting applications of computer cartography and a clear presentation of programming procedures.)

Burrough, P. A. (1986) *Principles of Geographical Information Systems for Land Resources Assessment. Monographs on Soil and Resources Survey* **12**. Clarendon Press. (Although essentially concerned with GIS, there is much about computer cartography.)

Carter, J. R. (1984) *Computer Mapping: Progress in the '80s. Resource Publications in Geography*. AAG, Washington. (A very good primer on computer mapping, with a north American orientation.)

DoE (1987) *Handling Geographic Information*. HMSO. (Particularly good on the sources of data and the applications of computer cartography.)

Forer, P. (1984) *Applied Apple Graphics*. Prentice-Hall, Englewood Cliffs, New Jersey. (A good book about microcomputer graphics in general and the Apple II in particular.)

Newman, W. M., Sproul, R. F. (1979) *Principles of Interactive Computer Graphics*. McGraw-Hill, Japan. (The basic principles of computer graphics for those who want a wider perspective. Getting a little dated.)

Robinson, A. H., Sale, R. D., Morrison, J. L., Muehrcke, P. C. (1984) *Elements of Cartography* 5th edn. Wiley, New York. (An excellent all round cartography book, with much about computer cartography.)

6

Remote sensing and image analysis

The term *remote sensing* was first coined in 1960 to refer to the observation of a target using a device located some distance away from it (Curran 1985). Since then it has developed discipline-dependent meanings. In the environmental sciences it generally means the use of Electromagnetic Radiation (EMR) sensors to record images of the environment which can be analysed to yield useful information. The term *image analysis* refers to the various computer methods used to process images in a digital format. The images may be aerial photographs, satellite scenes or even electron microscope micrographs.

Prior to 1925, remote sensing was little used by environmental scientists, but the potential utility of aerial photography was gradually recognized. From 1925–60 there was a great increase in the application of aerial photography, especially during the Second World War. Major developments were made in the techniques and applications of remote sensing for examining first spatial and then temporal variations in the environment. The period 1960 to the present has been easily the most active in the history of remote sensing. The launch of the first remote sensing satellite (TIROS–1) in 1960, heralded a phase of intense use of aircraft and satellites for collecting remote sensing data. In particular, the launch of the American Earth Resources Technology Satellite (ERTS–1, later renamed Landsat–1) in 1972, marked a new era in environmental remote sensing. Remote sensing and image analysis are now amongst the most important computer applications in geography and, indeed, in other disciplines such as astronomy, geology and zoology.

This chapter will first outline the basic principles of remote sensing and image analysis. There is then discussion about the data collection process, the nature of aerial photographs and satellite images, image analysis hardware and software, and the various operations involved in image analysis. Following this, the applications of remote sensing and image analysis are briefly considered and a case study of the use of remote sensing for mapping wetlands is presented.

THE PRINCIPLES OF REMOTE SENSING

Environmental remote sensing involves the use of sensors to record variations in Electromagnetic Radiation (EMR). The principal source of EMR is the sun, but the earth's atmosphere is responsible for filtering out certain wavelengths of the EMR spectrum (Fig. 6.1). Thus the most

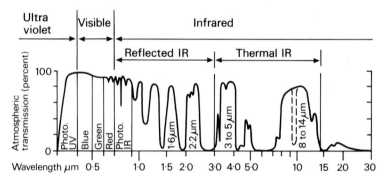

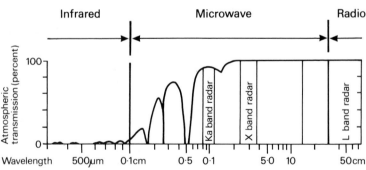

Fig. 6.1. Electromagnetic radiation from the sun which penetrates the earth's atmosphere. Note that the horizontal scale is non-linear and that the lower part follows on from the upper part.

Source: redrawn from Sabins 1987.

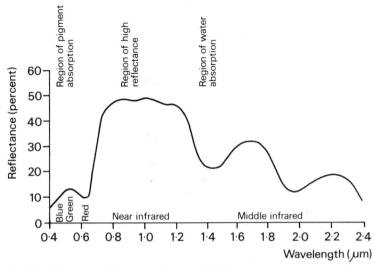

Fig. 6.2. The spectral signature of a green leaf.
Source: Curran 1985.

important wavelengths, as far as remote sensing is concerned, are visible and near (reflected) infrared (wavelength $0.4-3.0\,\mu m$), thermal infrared ($3-14\,\mu m$) and microwave ($5-500\,mm$). Virtually all objects in the environment emit and reflect both different quantities and types of EMR (their so-called spectral signature). Details about the emission and reflectance of objects can, therefore, be used to gain valuable information about the environment. The spectral signature of vegetation in summer, for example, derives principally from the reflectance, absorbance and transmittance of leaves (Fig. 6.2). Healthy green leaves have a characteristic spectral signature that has low reflectance of blue and red light, medium reflectance of green light and high reflectance of near infrared EMR (Curran 1985).

Remote sensing data can be collected, analysed and displayed as both *continuous* and *discrete* images. In continuous images, such as aerial photographs, the data are in analogue format. In discrete images, such as the majority of those obtained from satellites, the data are in digital format. Sophisticated techniques have been developed for processing both analogue and digital images. However, since the computers which geographers use are only capable of handling data in digital format, only discrete (digital) processing will be considered here. Analogue images can easily be converted into digital format using any one of a number of devices, such as a scanning densitometer or a video camera (see Jensen 1986).

Data collection

The key aspects of the remote sensing data collection process are shown in Fig. 6.3. The source of EMR may be the sun, the earth's emitted heat or a man-made source such as a power station. The amount and type of radiation emitted and/or reflected depends upon various characteristics of the environment, including the form of the earth's surface and the atmospheric conditions. The level of EMR is recorded by a *sensor* which is normally mounted on a *platform*.

The two main types of sensor are framing systems and scanning systems (Sabins 1987). Framing systems, such as cameras and vidicons (a type of video camera), instantaneously acquire an image. In both cameras and vidicons, a lens is used to gather light, which is passed through various filters and then focused on a flat photosensitive target. In a camera the target is a film emulsion and in a vidicon it is an electronically charged plate. Scanning systems employ a single detector with a narrow field of view, which sweeps across a scene in a series of parallel lines collecting data for contiguous cells to produce an image. Digital images derived from both these systems (either directly, or indirectly by analogue to digital conversion) comprise a series of cells called *pixels* (picture elements), each of which has a discrete intensity value (digital number) proportional to the level of EMR received by a sensor. The simplest types of sensors record a single image in a single spectral band. For many remote sensing applications, however, it is essential to record a scene in a series of spectral bands to give a

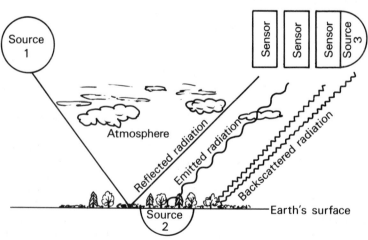

Fig. 6.3. The key features of the remote sensing data collection process. *Source*: Curran 1985.

multispectral image. Multispectral images may be recorded by aligning several cameras on the same scene, each with a different filter in front of the lens. This method has deficiencies, however, since each spectral band is collected through a different lens (which may give interpretational problems) and because the spectral range of cameras is restricted to visible and infrared bands. Therefore, virtually all multispectral images are collected using scanners.

Some sensors, for example many cameras, are hand-held and others are mounted on tripods or microscopes. However, many remote sensing applications require that a sensor is located at some distance from the earth's surface. In such cases, sensors must be mounted on a special platform, such as an aircraft, satellite, balloon, rocket or even a bird.

As far as environmental scientists are concerned, the most important types of images collected from remote sensing systems are aerial photographs and satellite images.

Aerial photographs

Aerial photography was the first major remote sensing technique and even today it is still widely used. All that is required is a light-proof box, a lens and a film coated with an emulsion sensitive to EMR (Drury 1987). Aerial photographs are widely available at a range of scales for much of the earth. They are much cheaper than field surveys and in many instances they are more accurate than maps. Because they offer a synoptic viewpoint, that is, a picture of a widely dispersed area at the same time and under the same conditions, they can be used to examine both spatial and temporal variations in the environment. Aerial photographs can be used to sense objects outside the spectral and spatial resolution of the human eye in the infrared area of the EMR spectrum. Overlapping pairs of aerial photographs, when viewed using a stereoscope, offer a three-dimensional perspective, something which is only rarely available from other remote sensing systems.

A variety of camera and film types are available for use in remote sensing. Their selection depends upon both their cost and the nature of the application. Ideally, cameras for aerial photography should have a high geometric accuracy, a medium to large format, a high quality lens, a suspension mount (to ensure that the camera is vertical) and a motor drive.

The types of film used for environmental remote sensing applications are classified according to their sensitivity to EMR. *Black and white* (also called panchromatic) film is the most widely used because of its geometrical stability, low cost of processing and printing, and high spatial resolution. *Black and white near infrared* film is similar in many respects to black and white film. Its greater spectral sensitivity means that near infrared wavelengths can be recorded in addition to visible light. This type of film is used for applications such as the identification of crop diseases and estimation of the soil moisture status of agricultural land. *Colour* film allows the human eye to distinguish many more features, because of its greater sensitivity to tints and shades of colour compared to tones of grey. Set against this, however, colour photographs are more expensive and have less good image definition. Colour film has a great number of applications. For example, its sensitivity to sub-surface water makes it especially useful for coastline definition and the estimation of water depth and sediment content. *False colour near infrared* film is similar to colour film, but the colour response of the film is changed to incorporate near infrared, as well as some visible wavelengths of EMR. This type of film has been used for applications such as detecting tree stress in forests and identifying crop marks on archaeological sites.

Satellite images

Satellite images have been collected using sensors mounted on satellite platforms since the early 1960s. Many types of satellite and sensor have been employed, but the basic principles are common to most systems, and so for convenience the Landsat–5 satellite (one of the most up to date earth resources satellites) will be used to illustrate the process of collecting satellite images. The key features of Landsat–5 are shown in Fig. 6.4. This satellite comprises two sensors, a MultiSpectral Scanner (MSS) and a Thematic Mapper (TM), together with associated devices for maintaining an anti-clockwise orbit over the poles and for handling data collection and transmission. The sensors on the satellite collect data by scanning parallel lines across the surface of the earth, in a direction perpendicular to that of the orbit. The rotation of the earth shifts successive orbit paths westward so that the whole of the globe is covered every 16 days. The information collected from a sensor is then transmitted to earth ground receiving stations. These data are available for purchase from a number of organizations on magnetic tape or on floppy disk.

The usefulness of satellites for remote sensing depends on a number of features (Fig. 6.5). Most of the satellites which collect data for use in environmental remote sensing, orbit the earth near the poles, obliquely or equatorially (for example Landsat–5 and SPOT–1 have a near polar orbit). The orbiting speed of these satellites is usually designed so that they are sun-synchronous; that is, they always collect data from areas in sunlight. These contrast with geostationary satellites which are fixed above some point on the earth's surface (for example Meteosat–2 is located on the Greenwich Meridian over West Africa). The repeat cycle is the rate at which the same area of the earth's surface is imaged by a

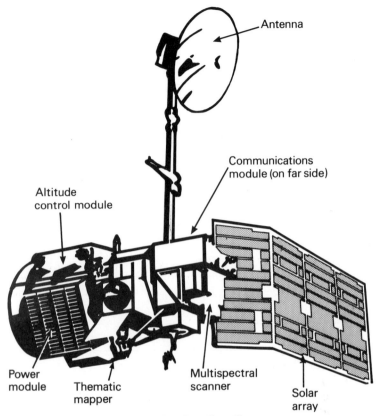

Fig. 6.4. The key features of the Landsat–5 satellite.

satellite sensor. Meteosat–2, which collects an image every 30 min, is an example of a satellite with a rapid repeat cycle and Landsat–5 is an example of a satellite with a much slower repeat cycle of 16 days. The spatial and spectral resolution of satellite sensors are important controls on what can be distinguished in the environment. The SPOT–1 satellite, with a maximum resolution of 10 m (that is, a pixel size of 10 × 10 m), is regarded as having a high spatial resolution. Towards the other end of the spectrum is Meteosat–2, with a maximum resolution of 2.4 km. The spectral resolution is determined by the number and size of specific wavebands and the possible range of intensity values. The thematic mapper sensor on Landsat–5 collects data in seven relatively narrow wavebands, whereas the push broom sensor on SPOT–1, working in panchromatic mode, collects data in only a single relatively wide band. The range of intensity values determines the number of discrete

	Landsat–5		*Spot–1*	
Orbit	Near polar		Near polar	
Repeat cycle	16 days		26 days	
Maximum spatial	MSS	TM	MS	Pan
resolution	56 × 79 m	30 × 30 m	20 × 20 m	10 × 10 m
Scene size	185 × 185 km		60 × 60 km	
Altitude	705 km		832 km	
Launch date	1/03/84		22/02/86	

	Meteosat–2
Orbit	Geostationary
Repeat cycle	30 minutes
Maximum spatial resolution	2.4 × 2.4 km
Scene size	¹/₄ earth's surface
Altitude	35900 km
Launch date	2/06/81

Landsat–5

Multispectral scanner		Thematic mapper	
Band width	Waveband name	Band width	Waveband name
0.5–0.6	Green	0.45– 0.52	Blue/green
0.6–0.7	Red	0.52– 0.60	Green
0.7–0.8	Near infrared	0.63– 0.69	Red
0.8–1.0	Near infrared	0.76– 0.90	Near infrared
		1.58– 1.75	Near-middle infrared
		10.4 –12.5	Thermal infrared
		2.08– 2.35	Middle infrared

SPOT–1

Multispectral mode		Panchromatic mode	
Band width	Waveband name	Band width	Waveband name
0.50–0.59	Green	0.51–0.73	Visible
0.61–0.68	Red		
0.79–0.89	Near infrared		

M e t e o s a t – 2

Band width	Waveband name
0.4– 1.1	Visible/near infrared
5.7– 7.1	Middle infrared
10.5–12.5	Thermal infrared

Fig. 6.5. The characteristics of three important remote sensing satellites.

units which can be allocated to a pixel. The multispectral scanner on Landsat–5 has a possible range of 64 intensity values. In comparison, the thematic mapper has the relatively high number of 256. The viewing area determines the size of the ground area which can be viewed at any one time. Meteosat images cover a ground area of approximately a quarter of the globe (Fig. 6.6a), Landsat images cover approximately 185 × 185 km (Figs 6.6b and 6.6c and Plate 2) and SPOT images cover approximately 60 × 60 km (Fig. 6.6d).

The physical constraints imposed by current technology mean that it is not possible to maximize each of these features. Depending upon the application for which a satellite is designed, choices need to be made between, for example, having a high spatial resolution and covering a large area with a rapid repeat cycle.

IMAGE ANALYSIS HARDWARE AND SOFTWARE

A digital image analysis system (henceforth the word digital will be omitted) is an integrated collection of computer hardware and software, which can be used to process data in the form of images. The data may be derived, not only from satellites, but also from radars, ground-based vidicon cameras and electron microscopes, amongst other devices. In theory an image analysis system is capable of processing any image that consists of a matrix of pixels containing digital numbers.

An image analysis system may be based on a mainframe computer, a minicomputer or a microcomputer. Jensen (1986) describes the operational characteristics of twenty-eight systems, which are either commercially available or are in the public domain, and Fig. 6.7 shows one commercially available system. The key hardware features of an image analysis system are: for *input* – a magnetic tape and/or floppy disk drive (for reading tapes/disks of purchased data), a keyboard and joystick or trackball; for *processing* – a host computer, preferably fitted with a coprocessor; for *storage* – a tape drive and/or floppy disk drive (usually the same one is used to read and store data); and for *output* – a high resolution screen for displaying images, a screen for text, a printer and a

(a)

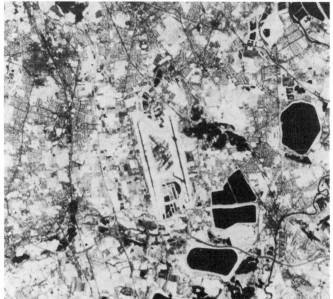

(b)

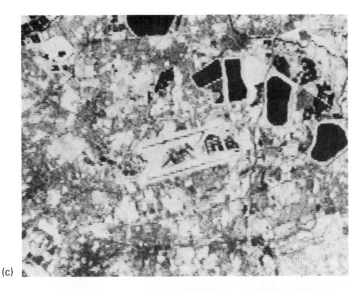

(c)

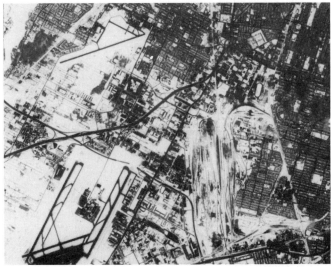

(d)

Fig. 6.6. Examples of satellite images. (a) A Meteosat visible and near infrared band image of the earth (provided by the National Remote Sensing Centre, Farnborough). (b) Part of a Landsat MSS band 4 image of Heathrow Airport, England at an approximate scale of 1:100,000 (provided by the National Remote Sensing Centre, Farnborough). (c) Part of a Landsat TM image of Heathrow Airport, England at an approximate scale of 1:100,000 (provided by the National Remote Sensing Centre, Farnborough). (d) Part of a SPOT-1 panchromatic mode image of Montreal airport at an approximate scale of 1:100,000 (provided by Nigel Press Associates, Edenbridge).

Fig. 6.7. A microcomputer-based image processing system (provided by Nigel Press Associates, Edenbridge).

hardcopy device capable of displaying images such as a camera or ink-jet plotter. All these devices are described elsewhere in this book; see especially Chapter 5 on Computer Cartography and Chapter 11 on Computer Hardware.

Jensen (1986) discusses the key factors which determine the suitability of an image analysis system for use in environmental remote sensing applications. The number of users a system allows is an obvious constraint. Most support only a single user, although some mainframe systems may allow limited time-sharing operations, and some minicomputer and microcomputer systems can be networked to share some hardware and software. The processor characteristics determine the speed of processing. Most images used by environmental scientists tend to contain a vast amount of data; a Landsat MSS image for example, is over 30 mb in size. It is important, therefore, that the processor can handle data efficiently and quickly. To facilitate this many systems are fitted with additional coprocessors. The operating system and compiler should be powerful, yet easy enough to use so that researchers can add their own programs and routines. The mass storage devices (tape and floppy disk drives) need to be able to deal with vast quantities of data. The display resolution may vary from only 40×40 pixels in microcomputer systems to over $1,024 \times 1,024$ in some powerful systems, although 512×512 is usually regarded as the preferred minimum for many applications. The colour resolution must be surprisingly large if images are to be displayed in full colour. A typical colour composite of three bands (the bands are displayed on top of each other) of a Landsat–5 thematic mapper image, for example, requires a single colour for each of the possible combinations of the 256 intensity values of each band, a total of 16,700,000 colours.

The workings of the hardware of many image analysers is largely transparent to the user, as a result of skilfully designed software. The software of some image analysis systems is amongst the most impressive and user friendly of any which geographers are likely to use. Many systems are menu-driven, that is, at each stage of the image analysis process, the user is prompted by an on-screen menu which lists all the possible options available. Default options are also normally programmed into the system (this means that the 'standard' option is selected unless the user chooses otherwise).

IMAGE ANALYSIS OPERATIONS

Image analysis systems should be capable of carrying out three main operations, namely, image pre-processing, image enhancement and image classification (Curran 1985).

Pre-processing

Pre-processing is the first operation to be performed on a newly acquired data set. It involves the restoration and correction of an image. *Restoration* includes the removal of any defect, such as the sixth-line banding effect on some Landsat images (brought about by variations in the response of scanners causing every sixth band to be either lighter or darker). *Correction* includes the suppression of effects such as atmospheric scatter (brought about by clouds and pollution) and the geometric transformation of an image to make it comparable with another map or image.

Enhancement

Enhancement involves the improvement of an image, using any one of a number of methods either singularly or in combination. *Contrast stretching* is used to improve the contrast of images by stretching out the range of digital numbers. In remote sensing this is often required because many sensors are tuned to collect data over very wide ranging conditions, from high reflectance areas like hot deserts, to low reflectance areas like the polar oceans. Consequently, only a small proportion of their measurement scale is used in any one scene. In multispectral images, *band-to-band ratioing* allows certain features to be enhanced. For example, subtraction of one wave band from another will result in features common to both being suppressed and other features will correspondingly be enhanced. This type of difference image can be used to monitor temporal changes, by comparing images of the same ground area taken at different times. *Spatial filtering* offers the ability to improve images by enhancing or suppressing certain spatial frequencies, directions or textures. One of the most commonly used spatial filtering techniques involves passing an array of N × N pixels (frequently 3 × 3) over an image and multiplying the digital number of the central pixel, by some function of all the values covered by the array. Depending on the weighting given to each element in the array filter, images can be smoothed and features in one or more directions can be enhanced. *Data compression* uses statistical techniques, such as principal components analysis, to reduce the data in several images, or bands of a multispectral image, to emphasize the key features. The resultant image can either be used for further analysis or for interpretational purposes. The final image enhancement technique is *colour display*. Since our eyes can perceive more colours than shades of grey, the use of colour can greatly enhance our ability to recognize features and patterns on an image.

Image classification

Image classification involves assigning pixels to categories on the basis of their digital number. The aim is to simplify an image by reducing the total range of digital numbers and spatial entities into the smallest meaningful number of categories. The two main methods of achieving this are a density slice of a single waveband or a supervised classification of several wavebands. *Density slicing* entails grouping together image regions with similar digital numbers. A single colour is then normally allocated to each class. This process is similar to classifying data in order to draw a choropleth map. The number and position of the class boundaries obviously has an important affect on the appearance of the final image. The details of the statistical procedures of *supervised classification* are quite complex. In simple terms a user identifies some areas on an image that are characteristic of each of the required classes. For example, in an attempt to derive a vegetation map from a scene of upland Britain, a user might identify areas of arable crops, grassland, heathland and woodland which have already been mapped using ground survey techniques. The image analysis system then attempts to classify each pixel in an image into one of these categories. Several different statistical methods are available for determining how to allocate pixels to categories and each has its own merits (Curran 1985).

Image interpretation

The most crucial part of any investigation which involves image analysis is, of course, the assessment of the quality of the analysis and the geographical interpretation of the output. An essential part of this process is ground survey. Ideally, a very large number of representative sample points should be mapped at the same time as an image is collected, but this is rarely possible because of time and cost constraints. The process of interpretation will clearly be greatly facilitated if the scientist performing the image analysis has a sound geographical training and is conversant with both the problem and area under study. There is insufficient space here to explore the process of image interpretation, though some brief details are given in the wetlands example presented later. For further discussion see especially Drury (1987), Lillisand and Kiefer (1987) and Sabbins (1987).

THE APPLICATIONS OF REMOTE SENSING AND IMAGE ANALYSIS

The applications of remote sensing and image analysis are legion. Even a list of all the subjects and areas of the earth in which these techniques have been applied would constitute a substantial section in this book. The aim of this section, therefore, is to present brief details of some examples of the applications of remote sensing and image analysis in four areas, namely, biogeography, geomorphology, climatology and meteorology, and human geography. A case study of the use of remote sensing for mapping wetlands will then be discussed.

In biogeography aerial photographs and satellite images have been used extensively to classify and map plant, animal and soil distributions. Using specific wavebands, the status of forestry and agricultural resources has been classified as healthy, diseased, vigorous, senescent etc. The Canadian Forest Service has developed techniques for collecting tree data from inaccessible locations, for economic and management purposes. They use large-scale aerial photographs (*c.* 1:2,500) in order to determine the tree type, age and production levels (Lo 1986). Since 1974 several US government agencies have collaborated to try to predict the area, yield and production of crops on a worldwide basis using satellite data. Two of the most well-known programmes are LACIE (Large Area Crop Inventory Experiment) and AgRISTARS (Agricultural and Resources Inventory Surveys Through Aerospace Remote Sensing). It is now possible to assess world crop production levels, for crops such as wheat, to within a few per cent of error (Lo 1986). Landsat MSS data has been successfully employed to delineate large scale soil boundaries which are related to climatic and vegetation changes. The boundaries are not observable on the ground or by using conventional aerial photography because of the restricted spatial coverage of these methods. Imhoff *et al.* (1982) were able to map the soils of arid and semi-arid rangeland in central Utah, USA using satellite data.

Geomorphologists use remote sensing for the detection, identification and mapping of earth surface and sub-surface features. Such information forms the basis from which to determine the processes at work. The large spatial and spectral range and multi-temporal nature of remote sensing data, make them ideal for studying elusive lithological and structural features, as well as process dynamics (Millington and Townshend 1987). In a classic time-lapse photography study, Stoffel and Stoffel (1980) used stereopairs of aerial photographs to record the eruption of the Mount St Helens volcano in Washington, USA. Remote sensing has been widely used in studies of snow and ice (Hall and Martinac 1985). The detection of sea ice in polar regions is important, because of its danger to shipping and oil exploration. In an attempt to try to understand the process of iceberg production, detailed studies using Landsat data have been undertaken along the western coast of Greenland, one of the major areas of world production. Many aspects of desert geomorphology have been examined using remote sensing. In the Mojave desert, California, USA, Suguira and Sabbins (1980) used radar imagery to map superficial deposits, such as sand-gravel fans, desert pavement and sand and playa deposits. Shih and Schwengerdt (1983) also used remote sensing data to derive a classification of arid geomorphic surfaces. Geomorphologists have in addition been major users of image analysis in microscope studies of soil thin sections and sediment particles.

In contrast to biogeography and geomorphology, where aerial photography is still of great importance, in climatology and meteorology most remote sensing data are derived from space platforms. This is primarily because of the much larger global and regional scales of interest. Both polar orbiting and geostationary satellites are used, and data are available for a wide range of spatial scales and spectral bands. Remote sensing data are commonly employed to assist in weather forecasting and images are regularly shown on television. Ball *et al.* (1979), for example, describe the development of thunderstorms over Britain using Meteosat−1 data. Visible and thermal infrared images have been extensively used to estimate cloud cover and rainfall patterns (Barrett and Martin 1981) and sea surface temperatures (Saunders *et al.* 1982).

Remote sensing has applications in human, as well as physical, geography. It has been widely used for producing land-use maps of urban areas. For example, Brown and Winer (1986) used high and low altitude aerial photographs to map urban vegetation cover in Los

Angeles, as part of a project investigating pollution levels. Aerial photographs have also been used to estimate population levels and the relationship between population size and land area. There have even been some attempts to delineate areas of urban poverty and to derive sociological indices for cities (Lo 1986).

The use of remote sensing for mapping wetlands

Jensen *et al.* (1986) and Jensen *et al.* (1987) describe the use of remote sensing data for mapping various aspects of the inland wetlands along the Savannah River, South Carolina, USA. Two aspects of their research work will be discussed here. The first involves the use of an aircraft multispectral scanner, thermal infrared data for mapping the effects of thermal water pollution on the wetlands. The second involves the use of Landsat MultiSpectral Scanner (MSS) and Thematic Mapper (TM) data to prepare a regional wetland map of the area.

The thermal infrared multispectral scanner data were used to identify and map the effects on the river basin of thermal effluent produced by the Savannah River Plant. The data were collected in the $8-14\,\mu m$, thermal infrared waveband, using an aircraft-borne scanner flying at 1,200 m, with a spatial resolution of 3×3 m and with 256 intensity levels. The best time to collect the data was thought to be in spring, just before dawn and after passage of a cold front. At this time there is a significant thermal contrast between land and water surfaces, the lack of sunlight eliminates the effect of shadows and solar heating, and there should be low humidity level (which should give good quality images).

The data from the sensor were first geometrically corrected for image distortion (due to sensor scanning variation) and to remove overlap of consecutive scan lines. The data were then standardized to the ambient river temperature, by determining the river temperature above the effluent source and then subtracting this value from the digital number of all pixels in the image. Next the image was density sliced and isolines were added to produce, what is in effect, an isotherm map (Fig. 6.8). The class boundaries chosen for the density slice were 0, 1, 2.8, 5 and $>10\,°C$ above ambient temperature. The image shows the effluent as a plume of warm water discharging from the source. Graphs of temperature transects across the plume have been used to assess the rate of dissipation down stream.

The use of Landsat MSS and TM imagery for wetland mapping was also evaluated as part of the project. Three Landsat MSS images obtained in spring 1977 were used to cover the area of the watershed. An image analysis system was employed to resample the images to give pixels that were 80×80 m square (rather than the normal 56×79 m) and mosaic the three images together. These data were then geometri-

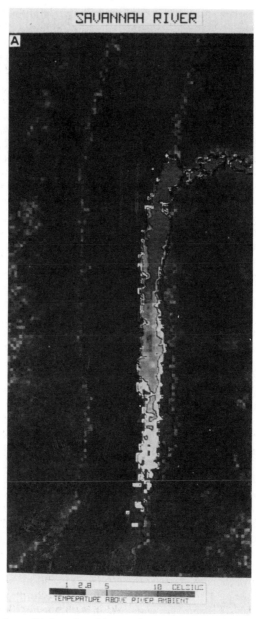

Fig. 6.8. A thermal infrared multispectral scanner image of thermal effluent pollution entering the Savannah River, South Carolina, USA, taken on 28 May 1983 (from Jensen *et al.* 1986).

cally transformed to the Universal Transverse Mercator Projection (an international standard map projection) to facilitate comparison with other types of data. A land cover map of the watershed was then created using a supervised classification of the image mosaic. Training areas were established for the nine standard US Geological Survey hydrologic units, using sample points which had already been surveyed on the ground. This procedure was found to be satisfactory for producing a map of the whole area (some 27,000 square kilometres), but was less useful for mapping small-scale features such as the effluent discharges described above.

The high spatial and spectral resolution of Landsat TM data suggests that in theory they should be eminently suitable for regional land cover mapping. However, at the time of the study the mechanical failure of the TM on Landsat–4 and only sporadic collection of Landsat–5, meant that the only image of the Savannah River region available for analysis was obtained on 28 August, 1982. Analysis demonstrated that, at this late date in the growing season, only using the near infrared and near-middle infrared bands was it possible to discriminate between the various wetland classes. The procedure used was similar to that described above for the MSS data. Jensen *et al.* (1986) conclude that a TM scene collected at some future date, which shows the area in the spring, may prove more suitable for land-cover mapping.

CONCLUSION

The fields of remote sensing and image analysis have expanded rapidly in the last ten years to become major applications of computers in geography. Remote sensing devices are responsible for the collection of massive amounts of environmental data. These data are now routinely processed using specialist 'turn-key' image analysis systems which combine 'state-of-the-art' computer hardware and software. The vast range of applications of remote sensing and image analysis in environmental mapping, monitoring and modelling, has led to their widespread use in many countries.

This chapter has briefly outlined the basic principles upon which remote sensing and image analysis are founded. The main characteristics of the hardware and software have been presented and some example applications have been discussed. At present remote sensing and image analysis are seen as clearly defined branches of geographical computing. In the near future, however, when the concepts and ideas involved in geographical information systems become more widely disseminated, remote sensing and image analysis should become more closely integrated with other areas of geographical computing.

FURTHER READING

Curran, P. J. (1985) *Principles of Remote Sensing*. Longman. (Companion volume to Lo (1986). A very good undergraduate introduction to the principles of remote sensing.)

Jensen, J. R. (1986) *Introductory Digital Image Processing: a Remote Sensing Perspective*. Prentice-Hall, Englewood Cliffs, New Jersey. (Good on image processing methods and details of image analysis systems.)

Lillisand, T. M., Kiefer, R. W. (1987) *Remote Sensing and Image Interpretation* 2nd edn. Wiley, New York. (A good straightforward up to date coverage for students.)

Lo, C. P. (1976) *Geographical Applications of Aerial Photographs*. David and Charles; Crane, Russak and Company Inc, New York. (Still one of the best books on aerial photography.)

Lo, C. P. (1986) *Applied Remote Sensing*. Longman. (A companion volume to Curran (1985). A good book on the applications of remote sensing.)

Sabbins, F. F. (1987) *Remote Sensing: Principles and Interpretation* 2nd edn. Freeman, New York. (A superbly produced book, with many colour illustrations.)

Short, N. M. (1982) *The Landsat Tutorial Workbook: Basics of Satellite Remote Sensing*. NASA, Washington DC. (All you ever wanted to know about the Landsat programme and satellite remote sensing.)

7

Simulation

Simulation is the process of designing a model of a real system and conducting experiments with the model for the purposes of describing, explaining and predicting the operation of the system. A number of different types of simulation model are normally recognized. *Scale* (also called hardware) models are physical models which are miniature copies of reality. For example, in a model of a valley glacier, clay may be used to simulate ice movement. *Conceptual* models are usually box and arrow (flow) diagrams drawn on paper which can be used to organize ideas. In *mathematical* models the key features of a system are translated into the symbolic logic of mathematics. All computer simulation models are mathematical models.

In a computer simulation model the salient features of a particular pattern or process are captured as a set of state variables. A procedure is then established which calculates a new set of values for the variables, taking into consideration the dynamics of the system under study. This procedure (also called an algorithm) is then translated into a computer program. During the operation of a computer simulation, the calculations indicated by the model's equations are performed repeatedly to represent change over space and/or time. The ability of a computer to perform these calculations quickly means that the effect of complex spatial and temporal problems can be considered.

The key features of a simulation are: the *input* – which might be rainfall values, population levels or rates of innovation diffusion; the *algorithm* – which is usually implemented as a series of equations; and the *output* – which might be tables of figures, maps or graphs.

The next section of this chapter will outline the nature of geographical computer simulation models (referred to from now on as simulations). The process of building a simulation is then outlined. Next the value and applications of simulations in geography are discussed. This is followed by four contrasting examples of simulations which illlustrate the main types of models and their uses in geographical research and teaching.

THE NATURE OF GEOGRAPHICAL SIMULATION

The development of computer simulation models in geography is one consequence of the quantitative and theoretical revolution (Walford 1981), which generally encouraged the use of systems theory, models and modelling techniques (Chorley and Haggett 1967; Minshull 1975; Kirkby *et al.* 1987). This concern with the dynamic elements of physical and human environments stimulated the use of simulation models and other similar techniques which offer more than inert snapshots of spatial systems. In this context simulation models may be classified as either deterministic or probabilistic (Thomas and Huggett 1980; Haines-Young and Petch 1986).

Deterministic models are based on the notion that the subject under study can be described exactly using mathematical relationships. Deterministic models are mainly, but not exclusively, restricted to physical geography where exact mechanical processes can be used to explain the behaviour of a spatial system. For example a deterministic simulation model of a drainage basin would deduce stream discharge (output) from rainfall (input). The nutrient cycling model discussed below is an example of a deterministic simulation model.

Probabilistic models are based on the notion that the subject under study has some random element. The equations of probabilistic models are based on the expected probabilities of certain events and processes occurring. For example, an epidemiological model of the spread of a disease through a population would be based on estimates of the probability of an infected and a susceptible person coming into contact. The factory location model discussed below is an example of a probabilistic simulation model.

It is perfectly possible for a simulation model to combine elements of both deterministic and probabilistic modelling strategies. The simulation of urban structure described below, for example, uses deterministic fractal recursion to create a number of areas. A probabilistic strategy is then used to choose the appropriate land-use type for the areas.

Many geographical computer simulations contain a *gaming* element. These introduce the ideas of competition and formal winners by structuring a simulation with explicit rules and by defining how a goal is to be achieved given certain resources. The oil slick simulation described below has a substantial gaming component.

THE MODEL BUILDING PROCESS

According to Roberts *et al.* (1983), the process of building a simulation model may be conveniently split into a number of phases (Fig. 7.1). The

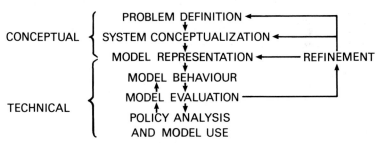

Fig. 7.1. The simulation model building process.
Source: redrawn from Roberts *et al.* 1983.

process begins with recognition and detection of a problem to study. The important influences believed to be operating on the problem system must then be committed to paper in the form of a conceptual model. This model is then represented in the form of computer code, using a general purpose programming language such as BASIC or FORTRAN, or else a specific purpose language designed to form and run computer simulation models such as DYNAMO (a contraction of the term DYNAmic MOdelling). The next stage is to run the model and ascertain how the variables in the system behave over space and/or time. Numerous tests are then performed on the model to evaluate its quality and validity. This might include checking for logical consistency, matching model output against observed data derived from real world systems (if available) and statistically testing the model parameters. The final phase of the model-building process involves using the model to simulate the system under study. It should be stressed that the process of building a simulation is not a simple orderly one which proceeds smoothly, but one which is characterized by much refinement (recycling and modification) of ideas.

113

THE VALUE OF COMPUTER SIMULATION MODELS

Simulation is a valuable method of geographical enquiry for a number of reasons. It is often cheaper to simulate the patterns and processes of inaccessible and distant places than to mount expeditions to test ideas in the field. Since it is not possible to construct physical models of the human decision-making process, computer-based mathematical simulations can be used to understand the behaviour of economic, social and urban spatial systems. Simulation allows the temporal dimension of processes to be compressed and extended, so long-term processes (such as population development and continental drift) and short-term processes (such as forest fires and land slips) can be conveniently investigated. Equally important is the fact that by repeatedly running a simulation, with key parameters set to different values, competing theories can be tested and evaluated in a relatively short time. For example, in the factory location simulation described below, weights are allocated to location factors such as government and local authority aid, residential desirability, accessibility and agglomeration diseconomies. By varying the weights of these factors each time the model is run their effects can be examined. Using simulation methods the possible effects of future catastrophes can be assessed. This might include simulating the effects of an earthquake on San Francisco, a volcanic eruption on Sicily, or a five-year drought on North Africa. Clearly it is not possible, not is it desirable, to build a scale model or have a trial run of such events under present-day conditions. Simulation modelling is also central to the idea that change is common to all human activity. As such, an understanding of the structures that cause change could be used as a cross-discipline organizing principle for studying the environment (Roberts *et al.* 1983).

Computer simulation is used for a number of valuable research and teaching purposes. It is used to describe geographical patterns and processes. This 'has a remarkable disciplining and clarifying effect, and increases the rigour and order in the thought processes, especially because concepts have to be sharpened and quantified whenever possible' (Reif 1973, p. 224). Simulation assists in the explanation of geographical patterns and processes by enabling the testing of hypotheses about the real world. It can also be used to predict future and past geographical distributions and spatial processes.

In a teaching environment simulation can be used to develop a range of student skills including graphicacy, numeracy, logical thinking and problem solving. Students can develop an understanding of a substantive area of geography and as a by-product they become further acquainted with information technology. A further valuable use of simulation is the teaching of decision-making processes using role playing.

THE APPLICATIONS OF COMPUTER SIMULATION MODELS

There are a host of applications of computer simulations in many branches of geography. The aim here is to present something of the flavour of these applications. It is impossible to be comprehensive, because of the large number and great range of current applications. The examples discussed below have been chosen to illustrate the great breadth of problems which have been considered by geographers. The first part comprises a brief review of the application of simulation models in selected branches of geography. This is followed by four case studies which illustrate the use of the main types of simulation models in different areas of geography.

Hydrologists have been using deterministic computer simulation models since the 1960s. Fleming (1975) describes the characteristics of nineteen computer simulation models used by agencies such as the US Army Corps of Engineers (for stream flow synthesis and reservoir regulation), the British Road Research Laboratory (for the design of storm drainage systems) and the US Department of Agriculture Hydrology Laboratory (for understanding the interaction between agricultural activities and hydrology). The use of a simulation model, for determining soil water conditions on hillslopes during drainage, has already been described in Chapter 1. Burt and Butcher (1986) also describe the use of a simple model designed for teaching purposes, that simulates aspects of hillslope hydrology. Their model concentrates on the levels of soil moisture storage and sub-surface run off, for different types of slope and soil properties.

In biogeography, simulation models have been used to study many aspects of ecosystem and community dynamics. For example, Moffatt (1986) uses a simulation model to study the process of eutrophication. The model was created using the DYNAMO language with the aim of understanding the relationship between rates of algal growth and nitrogen levels in natural waters. Barlow and Dixon (1980) also describe a simulation model of the population dynamics of the lime aphid which has been used to examine the importance of various control processes.

In climatology and meteorology, simulations have been used to investigate atmospheric processes and for weather prediction. For example, Dozier and Outcalt (1979) describe a simulation model of the energy balance of rugged terrain. Values for net radiation, soil heat flow, sensible latent heat flow and surface temperature were estimated for every point on a digitized grid and then output as contour plots. The model has applications for studying snow melt, water and heat flow in soils, and weathering. Pease (1987) uses a computer to generate a global energy budget model. The model is based upon a fixed level of energy input (short-wave radiation). This is redistributed according to the effects of various parameters that represent the state of the atmosphere and the earth's surface. Pease uses this model to estimate average global surface temperatures, to compare other published energy budget models, to test the sensitivity of surface temperature changes to energy budget parameters and to examine the temperature rise resulting from the addition of carbon dioxide to the atmosphere (greenhouse effect).

Simulation, of course, is not solely the preserve of physical geography. There are many examples of the use of computer simulation models in human geography. The urban dynamics model developed by Forrester (1969), for example, focuses on the dynamics of growth, stagnation and decline of a city. It considers changes in business, employment and homes. Using the model, the effects of policy decisions aimed at the recovery of a city may be investigated over a 250-year period. Probably the best known computer simulation model is the WORLD3 model discussed by Meadows *et al.* (1973). This analyses possible relationships between population, agriculture, resource use, industry and pollution at a world level. In the simulation, world trends over the 70-year period 1900–1970, are projected over the 30-year period 1970–2000. By changing various assumptions about rates of innovation, population change, government policy etc., various scenarios can be established about the state of the world.

A deterministic simulation of nutrient cycling

The nutrient cycling deterministic simulation was developed by Haines-Young (1983). It allows users to simulate the cycling of common soil nutrients (such as calcium, potassium, sodium and magnesium), under different climatic regimes and using different assumptions concerning the nature of the processes operating in ecosystems. The model comprises a series of compartments, in which a nutrient is temporarily stored, and a set of transfers which govern the scale and pattern of movement of a nutrient through an ecosystem (Fig. 7.2). The flux of nutrient between the soil (S), vegetation (V), and litter (L) stores is determined by transfers representing uptake (TV), litter fall (TL) and decay (TS). Nutrients may enter the system via precipitation input (PI) or from weathering (TM) of the nutrient in the mineral store (M). Leaching losses are represented by the transfer to drainage (TD).

The simulation is quick (a single run takes only a few minutes) and easy to run. Commands are entered from a menu and after screen

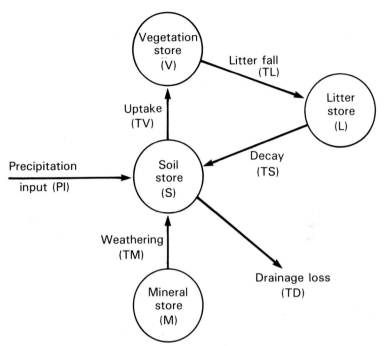

Fig. 7.2. The structure of the nutrient cycling simulation.
Source: Haines-Young 1983.

prompts. Output from the model is in the form of colour line graphs and tables of results (Figs 7.3 and 7.4). The graphs and tables show the size of the nutrient stores and the rates of transfer between the stores at the last step.

For the purposes of calibration and in order to increase user familiarity, the simulation may be run once using the data for calcium in the Hubbard Brook (New England, USA) forest ecosystem, which are provided. It is then possible to simulate the effects of various factors on the cycling of nutrients, using either the Hubbard Brook forest data or some new data supplied by the user. The simulation allows changes to two types of input. First, there are data relating to the initial conditions under which the simulation is to be run (for example, initial store sizes, climatic data and the mobility factors relating to the particular nutrient being considered). Second, there is a set of assumptions which govern the operation of the transfer functions and thus the movement of nutrients in the system.

The simulation, although simple in overall conception, has a number of uses. First, it may be used to focus attention on the methods of

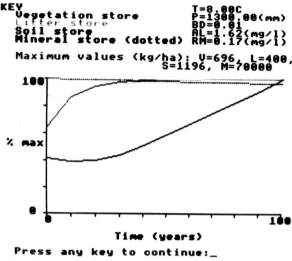

Fig. 7.3. Graphical output from the nutrient cycling simulation. The graphs show the cycling of calcium for the Hubbard Brook forest ecosystem.

formulating research questions and the need to develop models in geography. Second, it may be used to show how a simulation of nutrient cycling might be developed and tested. The contrast between traditional experimental methods and computer simulation methods has relevance here. Third, the simulation may be used to test hypotheses about the structure and operation of soil-vegetation systems. Since this is the main purpose for which the simulation was designed two examples will be considered in more detail below.

The first example concerns the effect of changing climatic inputs on the distribution of calcium in the simulated ecosystem. Using different combinations of mean annual precipitation and mean annual temperature it is possible to simulate the effects of 'moving' the Hubbard Brook ecosystem into other climatic regions. The data appropriate for 'moist-tropical', 'semi-arid', 'moist-temperate' and 'tundra' environments are given in Haines-Young (1983). The performance of the simulation can be assessed by comparing these simulated values with data from real world ecosystems.

The second example concerns the effect of changing the ionic mobility and biological demand for different ions. Together these parameters govern the rate of movement of a nutrient in an ecosystem. The biological demand (BD) is a measure of the amount of a nutrient in the standing crop of an ecosystem. Nutrients which have a high biological demand will tend to circulate faster. The weathering potential (WP)

T	V	L	S	M
0.	450.	350.	500.	70000.
10.	603.	323.	474.	69791.
20.	661.	361.	479.	69582.
30.	683.	384.	534.	69373.
40.	691.	394.	615.	69164.
50.	695.	398.	707.	68955.
60.	696.	400.	804.	68747.
70.	696.	400.	902.	68539.
80.	697.	400.	1000.	68331.
90.	697.	401.	1098.	68124.
100.	697.	401.	1196.	67917.

Press any key to continue

At the last step,
TV = 64.09Kg/Ha
TL = 64.09Kg/Ha
TS = 64.09Kg/Ha
TM = 20.69Kg/Ha
PI = 2.21Kg/Ha
TD = 13.11Kg/Ha
BD=0.01
AL=1.62
RM=0.17
Press any key to continue

Fig. 7.4. Tabular output from the nutrient cycling simulation. The tables
show the cycling of calcium for the Hubbard Brook forest ecosystem
(see text and Fig. 7.2 for explanation of terms).

describes the vulnerability of a nutrient to weathering. The relative
importance of these factors can be examined by comparing the cycling of
potassium with a WP of 0.23 and calcium with a WP of 1.65. Both have
similar biological demands. The model might be used to investigate
both the behaviour of potassium and calcium under different climatic
conditions and also which of the two ions is likely to be a limiting factor
for plant growth.

A probabilistic simulation of factory location

The factory location simulation may be used to examine a number of concepts of industrial location (Burcham and Ferguson 1985). In particular attention is focused on three aspects:

1. industrial location patterns are the result of combinations in inter-related factors;
2. some locational factors, for example a factory manager's image of places, may have an unpredictable or even random effect on factory location;
3. the location of a new factory may be affected by the location of existing factories (the multiplier effect or Weberian agglomeration principle).

The simulation is linked to the book by Bale (1981) and is briefly reviewed in Negus (1985).

The effects of industrial location on factory location may be investigated in any area. Since the simulation comes with the necessary data for a case study of eastern South Wales (Fig. 7.5), this will be used here. In eastern South Wales four factors are assumed to affect industrial location: government regional policy through financial grants to attract new factories; residential desirability; accessibility; and agglomeration diseconomies, which means that large cities such as Cardiff will generate

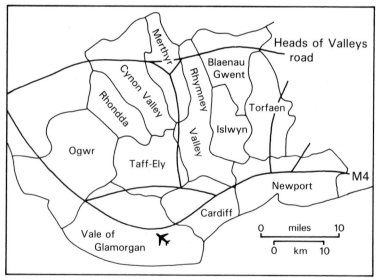

Fig. 7.5. The districts of eastern South Wales, the study area used in the factory location simulation.
Source: redrawn from Burcham and Ferguson 1985.

diseconomies (for example higher wages). The first three are positive factors which will attract industrialists, whereas the agglomeration diseconomies will tend to repel industrialists. The factory location simulation is a probabilistic simulation which uses Monte Carlo and Markov chain techniques. The term Monte Carlo has come to be associated with simulations that involve the generation and use of random or chance variables. In the factory location simulation, Monte Carlo processes are used to determine the probability of a factory being located in an area. Government aid, residential desirability and accessibility fall into this category. Markov chain techniqes can be used to model situations in which the probability of an occurrence is dependent upon an earlier occurrence. Thus in the factory location example the probability of a factory locating is dependent upon the number of factories already present. This enables consideration of the multiplier effect on factory location.

When the simulation is run a menu is drawn on the screen. This provides access to notes on the simulation, the data files (in this case the eastern South Wales data) and to the options for running the various parts of the simulation. During the first stage of the simulation it is possible to examine the effects of changes in government aid, residential desirability, accessibility and aggregation diseconomies, by allocating a weight to each of the factors based on their perceived relative importance. These global weights are then applied to the values in each of the grid squares which make up the region under study (the grid square values were established when the data file was originally created), to produce a probability matrix (Fig. 7.6). On this map the denser the shading, the greater the probability that a factory will locate in the area. Thus there is a high probability of factories locating in the Rhymney Valley (cf. Fig. 7.5). The computer can then be used to locate a number of factories in the region, say 100, on the basis of these probabilities.

During stage two of the simulation it is possible to use Markov chain processes to examine the influence of a multiplier effect on the location of factories. This simulates the influences which existing factories may have on the location of new factories. The multiplier may be positive – this will increase the probability of a factory locating close to an existing factory (simulating economies of agglomeration); or negative – this will decrease the probability of a new factory locating close to an existing factory (simulating diseconomies of agglomeration). It is possible to specify two agglomeration factors. The first applies a multiplier effect to those squares which already have a factory. The second extends the multiplier effect to squares adjacent to those already having a factory (Fig. 7.7). The effects of different multipliers can easily be examined by rerunning the model several times.

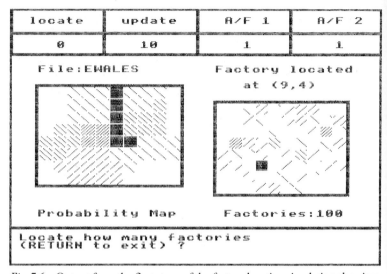

Fig. 7.6. Output from the first stage of the factory location simulation showing the basic probability map and location of 100 factories in eastern South Wales.

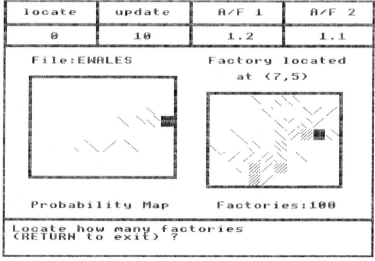

Fig. 7.7. Output from the second stage of the factory location simulation showing the basic probability map and the effect of two agglomeration factors on the location of 100 factories in eastern South Wales.

A deterministic/probabilistic simulation of urban structure

Urban geographers have long been concerned with modelling urban areas. They have tended to concentrate on economically orientated activities, expressed in terms of employment, population and transportation, at the macrospatial level in large zones, or at the microlevel of the individual or the firm. Batty and Longley (1986), however, chose a novel approach to simulating the physical configuration of land use itself. They used a computer simulation modelling technique based on the idea of fractal geometry.

The concept of fractals, which has recently become very fashionable, owes much to the work of Mandlebrot (1977). Fractal geometry is based on two tenets, namely fractal dimensionality and self similarity. The idea of fractal geometry cuts across the logic of classical geometry which classifies features using the integer dimension of length, that is, 0 for points, 1 for lines, 2 for areas, 3 for surfaces (see Chapter 2). Mandlebrot considers dimensions as a continuum. For example, a straight line on a map will have a fractal dimension of 1.0, an irregular line may have a value of 1.3, and a line which fills the whole page will have a value approaching 2.0. The idea of self similarity is that the complexity of feature form is independent of the scale at which it is viewed. For example, if a portion of a line is enlarged to the size of the original, the two lines, though different in detail, should show the same kind of overall complexity. Many features such as coastlines, administrative boundaries and land surfaces have been shown to exhibit these properties over certain scale ranges.

Batty and Longley's work on the simulation of urban land use seeks to reassert the idea that visual impact is as important as conventional parameter estimates and goodness of fit statistics in assessing the validity of geographical models. Thus the output from their simulations is displayed as maps which can be visually assessed. They simulate the location of three urban activities (commercial–industrial, residential and open space–recreational), using as a basis a classical von Thunentype model. A simple relationship of distance from the Central Business District (CBD) is used to establish probability profiles which give rise to concentric rings (Fig. 7.8). The simulation involves calculating the spatial distribution of land-use types, for different size land parcels (determined by fractal recursion), given the land-use profiles. The process of fractal enhancement is deterministic. It begins by dividing the original urban space into ten triangular sectors. Each sector is then subdivided hierarchically into more and more triangles based on the ideas of self-similarity. The triangles (land parcels) are then allocated land-use types. This is determined mainly by the distance from the

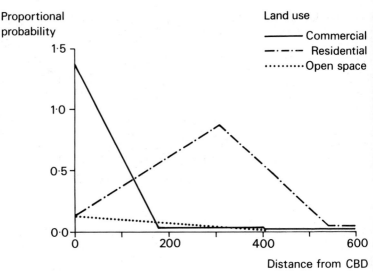

Fig. 7.8. Probability profiles for urban land use activities.
Source: redrawn from Batty and Longley 1986.

CBD and partly by the use of a random number generator within the computer program. The simulations at different levels of fractal recursion are shown in Fig.7.9. Up to recursion (r) level 2, the pictures are dominated by the original coarse triangular mesh used to generate the shapes of land-use activity. Above this level the structure appears much more satisfactory, although by level 5 the images become overcomplicated. Thus Batty and Longley (1986) conclude that levels 3 and 4 are the most appropriate simulations of land-use patterns in classical monocentric cities.

The use of fractals in simulation modelling has many applications beyond urban geography. Already fractals have been used to create landscapes for use in flight simulators and science fiction films (Batty 1987) and to simulate cartographic line detail (Maguire 1986b). Much experimentation into the potential use of fractals for describing and predicting other geographical phenomena has also been initiated.

A deterministic gaming simulation of coastal oil pollution

The coastal oil pollution deterministic gaming simulation was developed for the microcomputer by G. K. Blackwell, based on a game by M. Lynch (Cook 1984). The main aim of the simulation is to focus attention on the impact of technology on society through a case study of

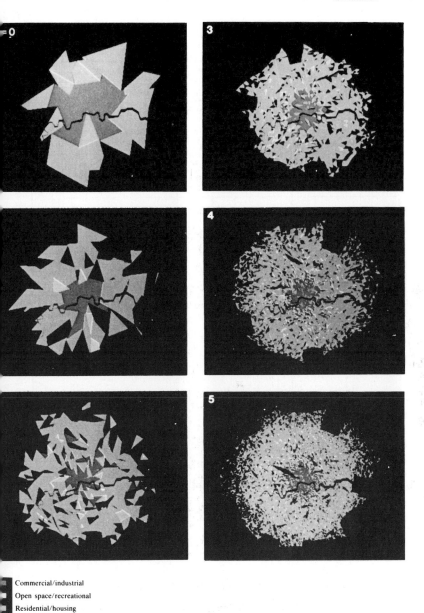

Commercial/industrial

Open space/recreational

Residential/housing

Fig. 7.9. Six simulations of urban land-use activity at different levels of fractal recursion.
Source: Batty and Longley 1986.

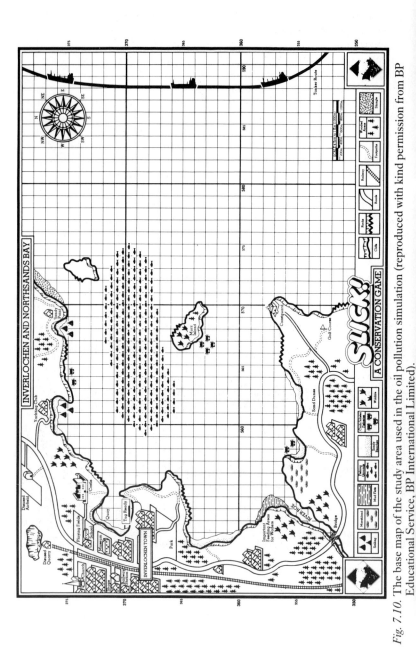

Fig. 7.10. The base map of the study area used in the oil pollution simulation (reproduced with kind permission from BP Educational Service, BP International Limited).

oil pollution and to investigate the modern means that are available to protect the environment.

When running the simulation, the user plays the role of local pollution officer in charge of an imaginary environmentally sensitive coastal area of north-east Scotland, called Northsands Bay (Fig. 7.10). This area is close to a major oil tanker route. The pollution officer has the task of minimizing the amount of environmental damage to the area by using a selection of anti-pollution methods (Fig. 7.11). These include materials to absorb, sink and disperse the oil out at sea, shore cleaning machines and a local task force. Booms are also available to try to stop oil getting ashore. The pollution officer has a limited budget and so only a selection of these methods may be purchased for use in any one run of the simulation. The first run utilizes the resources purchased by the previous pollution officer. The use of these resources is constrained by the speed at which they can be applied or moved and the potential environmental impact of their use. For example, it is inadvisable to apply sinker when oil slicks are above fishing grounds.

Before the start of the simulation the program presents a series of text and graphics screens with details about how to run the simulation (Plate 3). When the simulation is in progress the screen shows a map of the coastal zone (Fig. 7.12). Commands are entered via the keyboard, after

Fig. 7.12. The oil slick simulation in progress. The black square on the right hand side is the oil slick. A key to other features is given in Plate 3 and Fig. 7.10 (reproduced with kind permission from BP Educational Service, BP International Limited).

DATA SHEET

ABSORPTION 1

There are several materials which will absorb crude oil – straw, peat, polystyrene. These can be spread on an oil slick and mixed in. Then the oily mixture must be collected and disposed of on land either by burning or burial.

For 100 tonne spill

Time required (hrs)			Total cost
Prepare/ load	Travel time	Apply/ recover	
2.0	0.75	7.0	£2000

Advantages
- actually removes oil from surface instead of merely sinking or dispersing
- economic
- works well in calm, sheltered waters

Disadvantages
- its lightness makes it difficult to apply in high winds and seas
- difficult to collect – especially in bad weather
- suitable dumping places for disposal on land must be found

SINKING 2

Oil floating on the surface can be sunk by certain materials like specially treated chalk which floats until mixed with oil. It then becomes heavy and causes the oil to sink.

For 100 tonne spill

Time required (hrs)			Total cost
Prepare/ load	Travel time	Apply	
2.0	0.75	2.0	£1500

Advantages
- removes oil from surface of the sea eliminating danger to diving birds
- oil can be sunk quickly
- economic

Disadvantages
- oil is removed from surface but remains in the sea – may affect fishing grounds and fishing gear
- oil can still travel with underwater currents – may appear on shore weeks later if oil is sunk close to land

DISPERSANT 3

Dispersants help to split up oil. This allows natural forces such as waves, wind, sun and bacteria to break it down.

Dispersant is mixed with the oil after it has been sprayed on.

For 100 tonne spill

Time required (hrs)			Total cost
Prepare/ load	Travel time	Apply	
1.50	0.75	3.0	£3000

Advantages
- can be an effective, harmless method in the open sea

Disadvantages
- should not be used in harbours or close to shore inside estuaries
- should not be used close to shell fish beds for fear of contamination
- oil is not removed from the sea immediately

BOOMS 4

A boom is a floating barrier. Booms can be constructed from rubber, foam, wood etc. They can only be used close to the shore. They can be used in three ways:

1. As a barrier to protect harbours, rivers and other vulnerable areas from oil
2. To deflect oil from a sensitive area to a less sensitive area
3. To deflect oil into calmer waters

Details of a boom are given below. The boom is available in 20 m lengths which can be strung together to the desired length.

Lengths available	Cost per length	Time required (hrs)	
		Prepare/load	Install/anchor
20 m lengths	£50	1.0	2.0

NB: 1 square on the map measures 200 m.

SKIMMERS

When a boom has collected or deflected oil, the oil can be removed from the surface of water by a mechanical device called a *Skimmer*. This machine can collect oil at the rate of 100 tonnes per hour.

Skimmers can be hired from a firm located 30 miles north of Inverlochen. The cost of hire is £500 per day.

SHORE CLEANING 5

If oil does come ashore it will in time be broken down by the action of the sun, waves, and bacteria. But this cannot be allowed at the height of the holiday season!

Dispersants can be used and are sprayed along the beach before the tide comes in. Specially designed beach cleaning vehicles are available and could cover one typical beach in the Inverlochen area in 3–4 days. This would cost about £4000.

LOCAL TASK FORCE

Remnants of an oil spill most of which had been cleared at sea could be removed fairly quickly by a local task force. The task force would use straw to absorb incoming pockets of oil. Any oil on the beach would be removed by spraying with a low-toxic dispersant which would be hosed down with sea-water.

The cost of preparing such a local task force and having them stand-by to deal with a small amount of oil is £1000.

OIL AT SEA 6

Oil on the surface of the open sea moves in the direction of the prevailing wind. As it nears the shore it is usually affected by local currents. There are no strong local currents in Northsands Bay. Tides have a push-pull effect and can be ignored. So you have only to take account of the wind.

Details are given below of the effect of wind on an oil slick. Oil slicks move at about one thirtieth of the wind speed.

Wind description (Beaufort scale)	Wind speed (knots)	Number of squares slick will move on map every ¼ hour
Slight breeze	4–6	1
Gentle breeze	7–10	1
Moderate breeze	11–16	2
Fresh breeze	17–21	3
Strong breeze	22–27	3
High wind	28–33	4

Fig. 7.11. The various methods of dealing with oil slicks in the oil pollution simulation (reproduced with kind permission from BP Educational Service, BP International Limited).

screen prompts, and by selecting options from menus. The simulation runs for seven hours of simulation time (about twenty to thirty minutes real time) and the user is prompted for command input every fifteen minutes of simulation time. The simulation starts when a tanker is holed out to sea and an oil slick is blown towards the coast. In order to make use of the anti-pollution resources it is necessary to predict correctly the movement of the slick, based on the current wind speed and direction. While the simulation is in progress the computer keeps a record of the performance of the pollution officer. This is based on a number of criteria such as the number of correct predictions of the movement of the slick, the initial choice of resources, the amount of oil washed ashore and the impact of the slick on the marine environment (for example, points are deducted if dispersant or sinker are used over shellfish beds). At the end of the simulation the performance is assigned a numerical value that can be compared with a table of scores to assess the success of the pollution officer.

The present author has used the program successfully to teach second- and third-year geography undergraduates about the principles of simulation modelling and about coastal zone management. Practical use of the simulation often generates many questions such as:

1. How realistic is the simulation and how might it be improved?
2. What factors in addition to wind speed and direction might influence the movement of the oil slick?
3. How would the oil slick affect the inhabitants of Inverlochen (the harbour)?
4. What is the most effective method of dealing with oil pollution in coastal waters?

CONCLUSION

Computer simulation modelling is a technique that is increasingly being used by geographers from all branches of the discipline. Recent advances in computer technology, which have resulted in greatly increased computer power, have facilitated the increased use of computer simulation. Using current technology it is now possible to develop sophisticated models that contain many parameters. These can be used to augment often costly field-based studies and thus can help geographers in their attempts to describe and explain patterns and processes.

This chapter has provided a basic introduction to computer simulation modelling in geography. The key stages in the model-building process have been described, the basic types of simulation technique have been discussed and examples of the use of simulation in geography have been presented.

FURTHER READING

Ellison, D., Tunnicliffe Wilson J. C. (1984) *How to Write Simulations Using Microcomputers*. McGraw-Hill. (Some useful introductory remarks, guidelines and example routines for writing simulations on computers.)

King, R. (1981) (ed) Theme issue on games and simulations in geography teaching. *Journal of Geography in Higher Education* **5** (2). (The paper by Walford is especially useful and the other papers contain interesting examples.)

Kirkby, M. I., Burt, T. P., Naden, P. S., Butcher, D. P. (1987) *Computer Simulation in Physical Geography*. Wiley. (An interesting book on the use and programming of microcomputer simulation models in physical geography.)

Minshull, R. (1975) *An Introduction to Models in Geography*. Longman. (A relatively gentle introduction to models in geography with some discussion of simulation models.)

Neelamkavil, F. (1987) *Computer Simulation and Modelling*. Wiley. (A good introduction, but with few geographical examples.)

Roberts, N., Andersen, D., Deal, R. *et al*. (1983) *Introduction to Computer Simulation: a Dynamics Modelling Approach*. Addison-Wesley, Reading, Massachusetts. (A good all round introduction to simulation with some good examples.)

Word processing

Over the past five years or so, word processing has become one of the major computer applications. Although there is little explicitly geographical about word processing, it is one of a group of computer applications that have found favour amongst geographers, because it enables them to pursue their business more effectively and efficiently, and it is included here as such.

Word processing may be defined as the use of computers for the input, manipulation, storage and output of natural language. It is a part of the much larger field of text processing, which is concerned with information coded as characters or sequences of characters rather than numbers (Teskey 1982; Day 1984). Since there are few geographical applications of text processing at present, it will not be considered here. Note, however, that the project described in Chapter 1, on the computerization of Domesday Book, makes use of some text processing operations.

One of the most widely used word processing packages is WordStar. It is estimated that, since its release in 1979, over 1.5 million copies have been sold and it is now regarded as the *de facto* word processing standard. Today WordStar has been joined in the market place by over 250 other word processing packages (see Boggs Mathews and Mathews 1985 and Lang 1985c for general reviews). These figures clearly demonstrate the importance of word processing as a computer application.

The next section of this chapter will consider the benefits of using word processors. The section following this, on word processor hardware, includes discussion of printers. After this there is a section on word processor software, which includes a description of the basic facilities of word processing packages. The use of the WordStar word processing package for creating and editing a document is then described. Finally, there is brief discussion of Desk Top Publishing, a rapidly developing computer application, that offers considerable geographical potential.

THE BENEFITS OF WORD PROCESSING

Word processing has become so popular in recent years because of the facilities which it offers for interactive screen based editing, text formatting and advanced text manipulation. *Interactive screen based editing* may involve simply changing a single character or replacing a single word in a document. Alternatively, it may involve more sophisticated operations like replacing every occurrence of the same word in a long document of perhaps 20,000 words. For example, the 'English' English words in this book, such as colour, modelling and programmer could easily be changed to their 'American' English spelling (color, modeling and programer) using the global (that is change every occurrence) edit commands of a word processor. *Text formatting* includes such operations as justifying the left and right margins of text, changing the line spacing from single to double and altering the page length. Word processor commands for text formatting allow the layout and style of documents to be changed easily and quickly. Some word processors have facilities for *advanced text manipulation*. This includes such operations as generating indexes, checking spelling and merging standard letters with files of names and addresses (see below for further details of these operations).

Word processors can perform a number of valuable functions. They can be used for producing multiple drafts of documents, repetitive letters, small data files, and typewritten memos and notices. Many research papers, administrative reports and teaching documents need to pass through several drafts before they reach a final form. This may arise from the fact that some are of great complexity, whilst others need to represent committee views. As many such documents are of considerable length, word processors are an invaluable aid for making corrections and alterations. Probably the best known use of word processors is in the area of correspondence. This may entail production of a single letter to be sent to a single address or a standard letter to be sent to dozens of addresses. Multiple top copies of standard letters can be prepared for separate addresses by repeated editing or by merging the letter with a file of names and addresses. A lesser known use of word processors is the ability to handle files of information which frequently require updating. Reading lists, course handbooks and telephone directories are good examples of documents that can be more easily kept up to date when they are prepared and stored on a word processor. Word processors can also be used to produce neat typewritten memos, notices and labels, because all mistakes can be corrected interactively before the text is committed to paper. Even relatively slow two fingered typists, with illegible hand writing, can produce readable text without requiring the services of a secretary.

In summary, the advantages of word processors over manual text processing (pen and paper, or typewriter) are threefold. First, the accuracy and quality of typescript is improved. Second, productivity is increased and therefore in the long run costs are reduced. Third, access to information is improved and the storage space of files is reduced. Using these facilities work efficiency, at least in the long term, is considerably improved.

There are of course some problems associated with the use of word processors (Burges and Piercy 1985; Denley 1986). Few word processors have character sets with anything other than the English alphabet. This is a particular problem for many quantitative geographers, for example, who require Greek letters and Roman numerals for reproducing formulae. Word processors have a limited capacity. This is less of a problem than it used to be now that most computers have at least 512 kb of main memory and 360 kb of disk storage. In any case, long documents can be broken down into chapters or sections for separate processing. No matter how careful users are there will always be occasions when a whole document is accidentally deleted or a disk corrupted. But how many times have almost complete sheets of manual typewritten script been ruined? People who use word processors for extended periods often complain about their ergonomic design. Critics have particularly drawn attention to the problems of screen glare, screen height and angle, keyboard layout, noise and the possible harmful effects of radiation from screens. Lastly, the cost of word processors may initially be greater than that of typewriters. In addition to the financial costs, considerable time may be needed to learn how to use new hardware and software. On balance, it must be said that the advantages of increased efficiency, flexibility and quality, brought about by the use of word processors, more than compensate for these minor disadvantages.

WORD PROCESSING HARDWARE

This section will discuss the characteristics of the specialist hardware which is necessary for word processing, beginning with the processor. Although crucial for some applications, the type of processor is of only secondary importance in a word processor. This is because most word processing software will run on a range of machines. Similarly, the characteristics of processors important for other applications (such as the speed and size of data unit handled) are of lesser importance, since in word processing most of the time is taken up by data input and output. The users of word processors tend to be rather more concerned about the keyboard and screen. In particular, it seems that the 'feel' (sensitivity of the keys) of the keyboard and the 'look' of the screen are important. Some screens have white, green or bronze letters on a black background, whilst others have black letters on a 'paper white' background. The size of the screen and the number of columns and rows are also important. A primary requirement of a word processor is a large storage capacity. This includes both the amount of main memory and secondary floppy or hard disk storage. It is often surprising how much space text files occupy (remember that all the spaces in a file also count as characters) and so for large applications large capacity disks must be considered essential.

Printers

Printers are the basic hardcopy devices for alphanumerical data. As such they are frequently used in many applications, but for word processing a printer is essential. The main types of printer currently available are: dot matrix, daisywheel, ink-jet, thermal transfer and laser. These are discussed in turn below. The two main types of interfaces for linking printers to computers, the Centronics and RS232–C, are described in Chapter 9.

The most common type of printer, the *dot matrix*, fires a vertical line of metal pins onto an inked or carbon ribbon which then prints onto paper (Fig. 8.1). The shape of characters is determined by the selection of pins as the print head moves horizontally across the paper (usually printing in both directions). Modern ninepin print heads produce characters from a nine by nine matrix of dots. Dot matrix printers can print at high speeds of up to about 500 characters per second (cps), although most are in the range 100 – 200 cps. They can use both tractor-feed line flow paper (a continuous length of paper with perforations to mark the top and bottom of individual sheets and holes down both sides) and single sheet-feed stationery. Dot matrix printers are generally considered noisy, although recent machines are much quieter than their predecessors. They also tend to have a restricted character set, though some allow users to select from a number of different sets. A more serious limitation of dot matrix printers is their poor quality. Two methods of improving quality have recently been developed. First, extra pins have been built into the print head, increasing the total from 9 to 18 or 24, allowing printers to produce characters from a larger matrix of dots. These printers are generally equally fast, but are more expensive.

Fig. 8.1. Epson FX-80 dot matrix printer.

Second, some manufacturers improve print quality by making two, three or four passes over a character filling in the holes between dots. The penalty for this method is reduced speed, although it need only be used when higher quality output is required. Both these improvements aim to produce Near Letter Quality (NLQ) text, but they cannot rival the daisywheel printer for quality (Fig. 8.2). One additional advantage of dot matrix printers is their reliability; on the whole they require less servicing than other types of printers. Last but not least, an important consideration for geographers is the fact that dot matrix printers can be used to produce screen dumps of graphic images (see for example Figs 7.7, 5.7 and 5.8).

Daisywheel printers are very similar to modern typewriters. They operate by firing embossed characters onto an inked or carbon ribbon which in turn prints onto paper. The characters are mounted on the end of a multi-stemmed wheel which looks vaguely like a daisy, hence the name. Originally, daisy wheels were made of metal but more recently plastic has been used. The print quality is generally excellent and daisywheel printers can use both tractor-feed line flow paper and single sheet-feed stationery. On the other hand, daisywheel printers are noisy and slow. Typical speeds range from about 15 to 90 cps, depending primarily on the price of the printer. Lower price printers also have a

THE BRITISH ISLES

Position and size.—The British Isles comprise the two large islands of Great Britain and Ireland, together with a number of smaller islands, lying off the north-west coast of the continent of Europe. The English Channel, narrowing eastwards to the Strait of Dover, which is only twenty-one miles wide at the narrowest part, separates the south of England from France; the North Sea lies between Britain and Holland, Germany, Denmark, and Norway.

The island of Great Britain consists of the three countries, Scotland to the north, Wales in a part of the west, and England occupying the remainder. England, Scotland, and Wales have been under one king since 1603. Since 1920 Ireland has been divided into "Northern Ireland" and the "Irish Free State." Northern Ireland has a parliament of its own, but is otherwise closely united with Great Britain; but the Irish Free State is an independent dominion of the British Empire, having a president of its own.

"United Kingdom" used to mean the United Kingdom of Great Britain and Ireland now it means the United Kingdom of Great Britain and Northern Ireland. The distinction is important when comparing pre-war and post-war statistics, and care must always be taken when studying figures in official publications to see whether they apply to England and Wales only, or to Great Britain or to the United Kingdom.

Fig. 8.2. Comparison of output from different types of printer. Title and top paragraph – standard Epson FX-85 dot matrix printer; middle paragraph – Epson FX-85 Near Letter Quality (NLQ) mode; bottom paragraph – TEC F10 Daisywheel printer.
Source of original text: Stamp, L. D. 1938.

CENTRAL PLYMOUTH: OLD AND NEW *Bold and centre*

Brian S. Chalkley *centre* *Right justify*

⌐L

Introduction *Bold and underline*

This chapter seeks to provide a
commentary for a guided walk that
encompasses both the old and new city and
which underlines the contrasts between
the two areas. The tour, which allowing

PL

for discussion stops takes about 2 hours,
begins and ends at the city's main bus
station.

1. Bretonside Bus Station *Bold and underline*

This is only two minutes walk from the
Barbican, and so the vistor parties can
be conveniently dropped off here to begin
their tour. The name Bretonside recalls
an especially fierce French raid on the
10th August 1403. A force of 30 ships and
1200 men did considerable damage in the
area to the north of Sutton Harbour
before being repulsed.

Extract from:Maguire DJ, Brayshay M 1987
(Eds) Field excursions in the Plymouth *Underline*
region Straw Barnes Press, Retford

Fig. 8.3. The WordStar file Chalkley.1 after initial text input (see also Figure
8.4).

reputation for unreliability. One of the biggest drawbacks to daisywheel
printers is the limited range of characters which can be held on a single
daisywheel. Since this is usually only 96, multiple daisywheels must be
used for technical documents.

The market share of dot matrix and daisywheel printers is increasing-

CENTRAL PLYMOUTH: OLD AND NEW
Brian S. Chalkley

Introduction
This chapter seeks to provide a commentary for a guided walk that encompasses both the old and new city and which underlines the contrasts between the two areas. The tour, which allowing for discussion stops takes about 2 hours, begins and ends at the city's main bus station.

1. Bretonside Bus Station
This is only two minutes walk from the Barbican, and so the vistor parties can be conveniently dropped off here to begin their tour. The name Bretonside recalls an especially fierce French raid on the 10th August 1403. A force of 30 ships and 1200 men did considerable damage in the area to the north of Sutton Harbour before being repulsed.

Extract from: Maguire DJ, Brayshay M 1987 (Eds) Field excursions in the Plymouth region Straw Barnes Press, Retford

Fig. 8.4. The Wordstar file Chalkley.1 after editing (see also Figure 8.3).

ly being threatened by three new designs. *Ink-jet* printers work by shooting a fine stream of ink directly onto paper without using a ribbon. *Thermal* transfer printers melt dye from a special ribbon onto paper. *Laser* printers work in a similar fashion to photocopiers, but use a laser imaging system. They have very high resolution, typically, 300 dots per inch and are very fast; an A4 sheet can be printed in a few seconds (Figs 8.3 and 8.4 are examples of laser printer output). All these three types of printer can be used to print both text and graphics on the same page and the first two can print in both colour and black and white. At present they are little used in geography, but their future seems assured for certain specialist tasks. Laser printers, for example, are increasingly being used in Desk Top Publishing (see below) and ink-jet and thermal printers are used in remote sensing and image processing (see Chapter 6).

WORD PROCESSING SOFTWARE

By far the most important part of a word processor is the software. It determines the viability and ease of use of a word processor for specific applications. With over 250 packages currently on the market, this section will concentrate on the range of features available in word processing packages in general and will discuss the industry standard package, WordStar, in particular. Word processing packages each have their own command structure and syntax. In addition, they access the commands of the host computer operating system. Commands are entered either from a menu (menu orientated) or by the use of keywords (command orientated). Certain keywords, such as INSERT, DELETE and TAB, usually have specially identified keys.

The entry of text into a word processor is assisted by a process called wraparound. This means that when users attempt to type past the end of a line the software automatically carries the last incomplete word, or part of a word following a hyphen, over to the next line. A similar process called scrolling occurs when the bottom of a conventional 24 line screen is reached, that is, all the lines of text move up to leave a blank line at the bottom for text input. The screen is best thought of as a window showing part of a much larger document.

Editing

The editing facilities of word processors are invaluable. At its simplest, editing may involve using the DELETE key to delete the last character during text input. A number of more advanced commands are also provided for making revisions and corrections to documents after text input is complete. Movement around a document is facilitated by four keys which move the cursor up, down, left and right. Other keys allow movement to the top and bottom of a page, and start and end of a line. Changes can be made by typing new characters over old, deleting either a character, word, line, paragraph or page and inserting new text in its

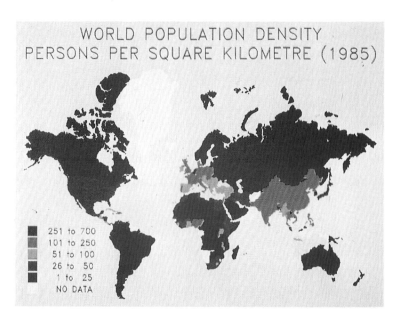

WORLD POPULATION DENSITY
PERSONS PER SQUARE KILOMETRE (1985)

- 251 to 700
- 101 to 250
- 51 to 100
- 26 to 50
- 1 to 25
- NO DATA

Plate 1

Plate 2

Plate 1 A GIMMS map showing world population density (see Chapter 5).

Plate 2 A Landsat–4 image of Mount Vesuvius. A quarter thematic mapper colour composite scene of bands 3, 5 and 4 taken on 24 January 1983 (see Chapter 6).

Plate 3

Plate 4

Plate 3 Output from the Slick simulation (see Chapter 7).
Plate 4 Output from the Domesday interactive videodisk system showing the distribution of soil types in Wales: Dark blue = sand, unripened gleys, man-made and lithomorphic soils (and sea), Light blue = pelosols and brown soils, Green = podzols, Yellow = gleys, Red = peats (see Chapter 10).

place. With the use of the INSERT key, new characters, words, lines, paragraphs and pages can be inserted. In addition, word processors can be used to move lines, paragraphs or pages of text from one part of a document to another (this is sometimes called 'cut and paste'). Many can also merge text stored in external files.

The majority of word processors have a range of more sophisticated editing commands, which allow users to search and replace characters or groups of characters called text strings. For example it may be desirable to search the text in Fig. 8.2 to locate the text string 'Germany'. After issuing the command, FIND Germany, the computer will search the text for the first occurrence of the string 'Germany'. Normally, the string is then highlighted by underlining it or printing it in reverse video. This part of the document can then be edited. The INSERT key could be used to insert five spaces before 'Germany' and the string 'West' could then be entered. Alternatively, most word processing software packages have facilities to search and replace text strings automatically. The last edit could have been completed by globally searching for every occurrence of the string 'Germany' and then replacing it with the string 'West Germany'. This is an extremely powerful function.

Formatting

The ability to format text is a further important feature of word processors. Most manual typewriters are designed to justify the left margin of text. Word processors can justify the left hand margin, the right hand margin or both at the same time. The latter is achieved using a technique called proportional spacing, which adjusts the size of the spaces between words. Some printers are also capable of microjustification. This involves adjusting the spaces between characters, according to the size of the character, so that narrow letters, such as 'i', take up less space than wide letters, such as 'm'. This gives a text format similar to that produced by a typesetter. The width of the page can also be set using these techniques by adjusting the position of the left and right margins. The length of each page is simply determined by fixing the number of lines per page. Other format commands exist for adjusting the spacing between lines, the size of top and bottom margins, the number of characters the start of paragraphs are to be indented and for automatically centring text. Two frequently used commands highlight text by printing it in bold or underlined. More advanced word processing packages also have provision for adding footnotes, subscripts and superscripts and for entering data in columns so that neat tables can easily be constructed. By using these commands repeatedly, users can experiment with the format of text before it is committed to paper.

Printing

When all the changes to the content and format have been completed, it is normal to print a paper (hard) copy of a document: the days of electronic mail and publishing are not yet the norm and paper remains the most important medium for exchange of the written word. One very important feature of word processors is that what appears on the screen should be exactly the same as that which appears on the printer. This property is sometimes referred to as WYSIWYG (What You See Is What You Get).

Most word processing packages offer some degree of software control over printing. Options usually exist for printing multiple copies of whole documents, pages or part pages. Most also allow printing directly from memory or disk. As printing can take a long time (perhaps a minute for a sheet of A4 paper on a daisywheel printer) many computer networks have special facilities for queuing documents for printers (see Chapter 9). The majority of word processors can drive more than one printer. This is achieved by including special routines in the software (called printer drivers), which utilize the range of fonts and special characters on specific printers. Some example fonts and character sets are shown in Fig. 8.5.

128	Underline
129	Bold (Double Strike)
130	*Italic*
131	Superscript
132	Subscript
133	Enlarged
134	Emphasized
135	Condensed
136	Elite
137	Proportional

```
138      Character set for USA
         # $ @ [ \ ] ^ ` { | } ~

139      Character set for France
         # $ à ° ç § ^ ` é ù è ¨

140      Character set for Germany
         # $ § Ä Ö Ü ^ ` ä ö ü ß

141      Character set for England
         £ $ @ [ \ ] ^ ` { | } ~

142      Character set for Denmark
         # $ @ Æ Ø Å ^ ` æ ø å ~

143      Character set for Sweden
         # Ö É Ä Ö Å Ü é ä ö à ü

144      Character set for Italy
         # $ @ ° \ é ^ ù ò à ò è ì

145      Character set for Spain
         ₧ $ @ ¡ Ñ ¿ ^ ` ¨ ñ } ~

146      Character set for Japan
         # $ @ [ ¥ ] ^ ` { | } ~
```

Fig. 8.5. An example of the fonts and character sets available using a word
processing package on an Epson FX-85 dot matrix printer.

Macros

Most word processing packages have facilities for creating user-defined
macros. These are collections of text and commands which can be used
for repetitive non-standard operations, such as the printing of personalized standard letters that have names and addresses merged from other
files. Macros are similar in concept to subroutines used in BASIC and
FORTRAN programming (see Chapter 12).

Accessories

Most packages have a number of optional extras including spellcheckers, mail mergers, indexers, sorters, grammar checkers and interfaces to

other packages. These may be sold either by the word processing package developer or by separate software companies. More advanced word processing packages may themselves contain some of these facilities.

Spellcheckers search through a document checking each word against a pre-defined dictionary of words. If the spellchecker finds a word not in the dictionary the word is identified as a possible error. Users are then normally presented with a number of options. If the word is spelt correctly, but is not in the dictionary, the user can decide whether or not to add the word to the dictionary. If the word is spelt incorrectly the user can correct the spelling. If the user is unsure about the spelling the dictionary can be searched to identify the words with the closest spelling. New dictionaries for technical and foreign words can easily be created. Spellcheckers, however, are far from perfect. They cannot be used to locate correctly spelt words which are misused, or are in the wrong tense. Small dictionaries frequently highlight correctly spelt words simply because they are not in the dictionary. Most dictionaries contain a minimum of 25,000 words and large dictionaries may hold 170,000 words.

Mail merging allows personalized standard letters to be sent to a number of people. This might involve sending details of a conference to all the people on a study group mailing list, or sending a report to several sponsors. The exact procedure varies from package to package, but generally it involves setting up a file containing a standard letter (for example conference details), with special characters embedded at the point where data in the mailing list file are to be inserted (for example name and address). When the job is run the two files are combined together to produce a standard letter for each person on the mailing list.

Indexers are used to create page indexes for reports and books (like the index at the back of this book). *Sorters* can be used to sort files containing alphanumerical data, such as names and addresses, into alphabetic ascending or descending order. *Grammar checkers* operate rather like spellcheckers. They perform to a set of pre-established rules and check for overlong sentences, repetitive phrases, punctuation and capitals. Unfortunately, their use is fairly labour intensive because of the great variety of language styles. Most word processing packages also have the facility to transfer data to and from other packages, such as spreadsheets, data base management packages and graphics packages.

CREATING AND EDITING A DOCUMENT USING WORDSTAR

MicroPro's WordStar has been the market leader for several years and even though it has a number of deficiencies it is now accepted as the *de facto* standard word processing package. It is available for most micro-computers which can run one of the standard computer operating systems and have at least 64 kb of memory. Although some commands are chosen from a menu, WordStar is basically a command-orientated package. The majority of commands are evoked by pressing either a single alphabetic key or by pressing the CTRL key with one or more alphabetic keys.

A WordStar session is initiated by typing WS at the operating system prompt followed by a press of the return key. The opening WordStar menu is then displayed on the screen. This presents a series of options such as *Open* a document file, *Copy* a file, *Delete* a file and *Print* a file. To create a new document the *Open* a document option is selected from the menu. A suitable name for the file, such as Chalkley.1, must then be entered. The WordStar menu is now replaced by a screen which is blank except for the top two lines. These contain information about the current status of WordStar (for example, the name of the file being created/edited). Any text now entered will appear on the screen starting in the top left hand corner where the screen cursor is located. Fig. 8.3 shows the document Chalkley.1 which was entered into WordStar in this way. Where the text to be entered occupies more than a line the process called wordwrap automatically carries the last incomplete word onto the next line and allows continual input. In situations where there is less than a complete line of text (as in the case of the headings and the ends of paragraphs) a press of the RETURN key is used to move the cursor onto the next line.

Once all the required text has been entered it is possible to edit the file to correct mistakes and to improve the lay out of the text (this could have been done during text input but for instructional purposes it is described separately here). A number of suggested changes are shown in freehand on Fig. 8.3. The two spelling mistakes are easily corrected. In the case of Chalk;ey, the cursor key is moved to the letter k and the offending ; is deleted using the command ^ G (a press of the CTRL key followed by the G key), insert mode is selected using the command ^ V and the new character l is entered (Fig. 8.4). To print a string of characters in bold the cursor is moved on top of the first character in the string and the commands ^ V (insert mode) and ^ PB are entered. The cursor is then moved to the end of the text string and ^ PB is entered again. Because WordStar is not truly WYSIWYG the string does not appear in bold but is bracketed by the characters ∧B on the screen. The process of underlining is exactly the same except that the command ∧PS is used instead of ∧PB. Centring is achieved by placing the cursor at the end of the string to be centred and then entering ∧OC. The text is right justified by first setting the position of the right margin using ∧OR and then reformatting each of the paragraphs using ∧B. The format of a document can also be controlled by a number of 'dot commands', so called because they begin with a dot. For example, the command to set the length of a page to 40 lines is .PL 40.

Once all the necessary changes have been made to the document a copy can be saved on disk by entering the command ∧KD. To print a copy of the document on paper the *Print* option is chosen from the opening menu. To end a WordStar session the *Exit* option is chosen from the opening menu. Help on the syntax and effect of commands is always available by entering the command ∧J.

The biggest single advantage of WordStar is its widespread use. It is virtually error-free, it has extensive good quality documentation, on-line help facilities and is well supported from the point of training and user enquiries. It has a good range of commands for highlighting text (including superscripts, subscripts, changes of pitch and changes of ribbon colour which could not be discussed above because of space constraints). It is possible to edit and print documents at the same time. One particularly useful feature is that WordStar automatically creates a backup copy of files. WordStar probably has the best selection of accessories of any word processing package.

On the debit side, WordStar is only partially menu-driven and more importantly it is not fully WYSIWYG. It lacks page-oriented commands, that is, it is difficult to move to the top or bottom of a page or on to a specific page and it is not possible to obtain multiple print copies of files. Additional problems include the fact that WordStar does not have a macro facility for creating user-defined command routines and that it is also rather expensive. A number of these problems have been resolved in more recent versions such as WordStar 1512 and WordStar Professional. WordStar is reviewed in Ettlin (1982), Boggs Mathews and Mathews (1985) and McTaggart (1985).

DESK TOP PUBLISHING

Desk Top Publishing is a recent computing application which is an entirely microcomputer phenomenon. It is based on two developments, namely, increasingly powerful graphics-orientated microcomputers and laser printers (although much lower cost dot matrix printers can be used for budget systems). It is the combination of these two pieces of hardware, together with an appropriate software package, which has made Desk Top Publishing such a rapidly developing application (Hammond and Stobie 1987). A Desk Top Publishing system is basically a combined graphic design and word processing system, that allows a wide selection of graphic images and character sets and fonts to be combined easily and quickly on the same page. It is easy to foresee many uses for Desk Top Publishing systems, especially in a subject like geography which is graphically-orientated. For example, a Desk Top Publishing system could be used to produce many documents, including advertising leaflets and posters, manuals and reports. The computer is responsible for typesetting, page design and make up and the laser printer gives very rapid, good quality output of both text and graphics.

At the heart of a Desk Top Publishing software package is a page description language. This is really just a powerful, high level printer control language that gives the user full control over what appears where on a printed page. All the standard word processing commands described earlier in this chapter are included, together with additional commands, for such operations as importing graphic images from graphics packages, producing multicolumn pages (like newspapers) and changing character fonts.

CONCLUSION

The aim of this chapter has been to demonstrate the great utility of word processing to the geographer. It can make a valuable contribution to several aspects of the geographer's work including preparing research papers, correspondence, reading lists and course documents. The important characteristics of word processing hardware and software have been discussed and an example of using the industry standard package WordStar has been described. There was also brief discussion of Desk Top Publishing. In the past few years there have been considerable advances in word processing hardware and software. The trend is towards easier to use systems which are menu-orientated, are fully WYSIWYG and can run on a wide range of computers, using any one of a large number of printers. There is also a move toward more integrated systems which combine basic word processing facilities along with a spellchecker and mail merger.

FURTHER READING

Boggs Mathews, C., Mathews, M. S. (1985) *Word Processing for the IBM PC & PCjr and Compatible Computers*. McGraw-Hill, New

York. (Good IBM coverage and software reviews of major packages.)

Ettlin, W. A. (1982) *WordStar Made Easy* 2nd edn. Osborne/ McGraw-Hill, Berkley. (A useful introduction to using WordStar.)

Letcher, P. (1985) *Which Peripherals? How to Choose Them, How to Use Them*. Chapman and Hall/Methuen. (A good readable book. See Chapter 3 for details of printers.)

Morgan, R., Wood, B. (1982) *Word Processing*. Longman. (Primarily concerned with the principles of word processing. Aimed mainly at managers and secretaries.)

Rosen, A. (1983) *Getting the Most out of Your Word Processor*. Prentice-Hall, Englewood Cliffs, New Jersey. (Good on hardware characteristics.)

Communication

Electrical communication links are required between computers and other machines for many purposes. These include the collection of data using data loggers, the transfer of data between computers, the access of external data bases and the control of external devices such as plotters and printers. Many geographical computer applications are, therefore, predicated upon good communication links. Some of these have already been discussed and others will be discussed below. The process of electrical communication between machines comprises a number of stages. In the simple case of data transfer from, say, a computer to a printer, the data in the memory of the computer are passed out through a communications interface port (normally a plug socket on the back of a computer), along a connection link (normally a plastic-coated wire), in through the printer communications interface port and into the memory of the printer. Many different types of link are used to connect machines together, but the same basic principles are common to all forms of electrical communication.

The next section of this chapter will outline the basic principles of digital communication. It includes brief discussion of analogue to digital conversion, data transmission patterns and communication interfaces. This is followed by consideration of the various types of communication link between machines, namely, direct lines, networks and telephone links. Examples of local and wide area networks are presented. Videotex, a major computer application which is dependent on good communication links is then discussed. Finally, there is consideration of the geographical applications of the Prestel viewdata system. This chapter contains a number of technical terms which are more fully explained in Chapter 11 and in the Glossary.

THE PRINCIPLES OF DIGITAL COMMUNICATION

Digital computer communication is, except in a few cases, relatively simple. Nevertheless, it is a difficult subject to master, because of the lack of universal *protocols* (communication conventions) arising from the fact that each of the many equipment manufacturers has different communication standards for each piece of equipment.

Analogue to digital conversion

The items of equipment which geographers use handle either analogue or digital data. However, since the computers which geographers use only handle digital data, the connection of computers to analogue machines requires analogue to digital and/or digital to analogue data conversion. Analogue data collection (input) devices that may be connected to digital computers include solarimeters (that measure solar radiation) and flow meters (see Chapter 2). Analogue output devices include plotters and videodisk players (see Chapters 5 and 10). Since digital to analogue conversion is, in principle, just the reverse of analogue to digital conversion, only the latter will be considered here.

Analogue to digital converters (also called A/D converters) take an analogue input (a current or voltage) and convert it into an equivalent digital number in binary form (a collection of 0s and 1s). The digitizing of an analogue signal will, in most cases, result in loss of information. The loss is proportional to the range of the analogue signal and the number of digits used to code the output. Typical A/D converters used in geography can deal with analogue input ranges of 0–2.5 V to 0–20 V and have 8 digit output. The output can, therefore, have 2^8 (= 256) possible values. The resolution of a 0–2.5 V analogue input, 8 digit output is, therefore, $2.5/256 = 0.0098$ V $= 0.98$ per cent. By contrast a good quality chart recorder might have an accuracy of 0.1 per cent. Clearly the measurement resolution of any data collection system is limited by the piece of equipment with the lowest resolution. There is little point, therefore, in connecting a high resolution A/D converter to a low resolution data collection device, since the overall resolution will be limited by the data collector. Not only may it give a false impression of accuracy, but it is also inefficient since A/D converters which output data coded using large numbers of digits (at present that is greater than 16) are expensive.

The sampling interval of an A/D converter should be based upon the time scale of the event in question and the amount of storage capacity available. If the period between samples is too short, too much data will be collected. On the other hand, if the time period between samples is too long, the analogue signal will be degraded and the phenomenon of aliasing may occur. Figure 9.1 shows how a true signal (dotted line) is misrepresented when samples (black dots) are too far apart. The solid line shows the apparent trend output from the A/D converter, which although erroneous may, superficially, appear correct.

The operation of an A/D converter may be affected by signal noise. The voltage which is supplied as input to an A/D converter will not be absolutely steady, but will contain some random and systematic fluctuations. These may result from electrical interference or mechanical vibration. Special software with appropriate filters may be required to eliminate such noise.

Transmission patterns

Digital data, in the form of binary 0 or 1 can be transmitted either several bits at a time using a number of adjacent lines (in parallel) or one bit at a time along a single line (in serial). To minimize error and maximize

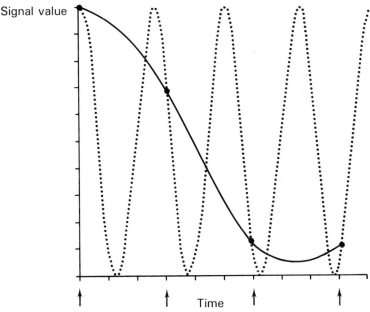

Fig. 9.1. The phenomenon of aliasing.

speeds, protocols (such as error checking and standard connections) have been established for both parallel and serial transmission.

The *parallel transmission* of data is comparatively rapid, but it can normally only be used over distances less than 2–3 m. This is because of the high cost of cable and the problems of interference between adjacent lines and data skew (some lines transmit faster than others, due to wire thickness etc., and so some data bits reach the receiver before others). In computing, the main parallel interfaces are the IEEE 488 and the Centronics printer interface. The Centronics interface is still commonly used to connect printers to computers and the line designation for the interface is shown in Fig. 9.2. Lines 2–9 are used for transmitting 8 bits of data and lines 1 and 10 are used for control. All the other lines are connected together to form the signal ground, which acts as a reference for the signal voltage. Data transfer using parallel communication is 'byte serial, bit parallel', each bit (an individual digit) of a byte (a group of 8 bits) is transmitted in parallel, but each byte is transmitted in serial.

Data transfers in *serial transmission* are 'byte serial, bit serial'. This is considerably more economical because transfers can be accomplished using a minimum of only two lines. One line is used for the data, which is transmitted as an electric charge (signal line), the other is used as a negative to complete the electric circuit (ground line). Serial transmission can take place over much greater distances than parallel transmission, because there is little chance of interference between lines. It also has the advantage of being widely used in telecommunications. This means that many devices already use this pattern of transmission and

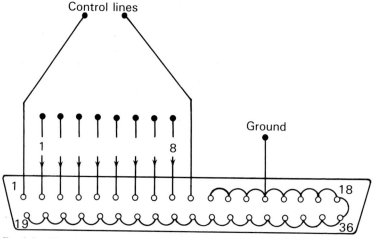

Fig. 9.2. The pin assignments used in the Centronics standard interface.

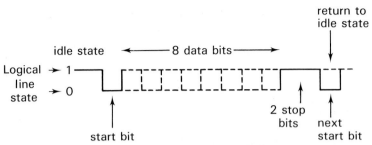

Fig. 9.3. The bit pattern required to transmit a single character asynchronously.

that the cost of components is relatively cheap. The disadvantage of serial transmission is that it is relatively slow, because data are transmitted only one bit at a time. In serial transmission some form of data transfer control is often required to indicate the end of one group of bits and the start of another. Both asynchronous and synchronous serial transmission protocols have been established for this purpose.

Asynchronous transmission is achieved by the addition of start and stop bits to the data bits (Fig. 9.3). The number of start bits is normally fixed at 1, but 1 or 2 stop bits may be added to the data bits before the sender transmits the data. The start bit is normally low (0) and the stop bit(s) high (1). The function of the start bit is to start a clock which will sample the signal line a given number of times (usually eight), at a frequency determined by the speed of the transmission (see below). The stop bit(s) allows the receiver to return to such a state in which it can recognize a new start bit. A new start bit can occur immediately after the end of the stop bit or after any period of time. Asynchronous transmission uses only two lines and can therefore easily be used for transmission down telephone lines. However, both the sender and receiver require a clock so that the data bits can be placed on the line at a fixed time interval.

In *synchronous transmission* a single clock is used to control transmission. In situations where the sender and receiver are quite close, a common clock signal is sent down a special line alongside the signal line. Various types of clock signal may be used, though a wave signal with a fixed frequency is most common. Synchronous transmission is more efficient than asynchronous transmission, since start/stop bits are not used and 8 rather than 10–12 bits are, therefore, transmitted per data unit.

RS232–C interface

The most common form of communication interface protocol used in computer systems is the RS232–C and its derivatives. These interfaces

Fig. 9.4. DB25-pin plugs which are used as connectors in the RS232-C standard interface.

normally utilize DB25-pin plugs for connectors (Fig. 9.4), although some manufacturers prefer 5-pin DIN plugs. In the protocol specifications, 21 of the 25 lines are defined, but not all of them are normally used for computer communication. A list of commonly used pin assignments is shown in Fig. 9.5.

Pin	Name	Input/output	Function
1	Protective ground	–	Maintains chassis of both devices at same potential
2	Transmitted data	Output	Signal line
3	Received data	Input	Signal line
4	Request to send	Output	Handshake line
5	Clear to send	Input	Handshake line
6	Data set ready	Input	Handshake line
7	Signal ground	–	Reference for signal voltages
20	Data terminal ready	Output	Handshake line

Fig. 9.5. Commonly used pin assignments in the RS232–C standard interface.

Handshaking

Handshaking is a widely used communication protocol for controlling both parallel and serial data transmission. An extra line, in addition to the signal and ground line, is normally required to carry the handshaking signal. A handshaking sequence begins by the sender signalling on the handshaking line (strobe) that data are about to be sent (Fig. 9.6).

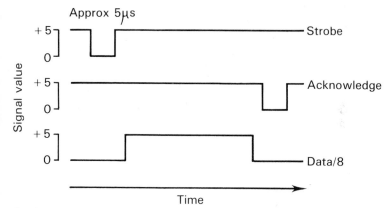

Fig. 9.6. The handshaking sequence for data transfer according to the Centronics standard. Note that only one of the data lines is shown.

The signal may take the form of a drop in the current from 5 V to 0 V for 5 ms. The data are then sent down the line. The receiver acknowledges receipt of the data by a signal on the handshaking line. Handshaking is widely used because it allows data transmission at the highest speed permitted by the slower of the two devices.

Error parity

In spite of the data transmission protocols outlined above, some errors may occur in data during transmission. Several error detection schemes are used to combat this problem, the most popular of which is the *parity check*. This uses an extra bit transmitted immediately after the data. This is set to 1 if the number represented by the transmitted binary code is odd or 0 if it is even. A check is then performed by the receiver to see if the parity bit matches the transmitted binary code.

Transmission speeds: baud rate

The rate at which characters can be sent along a communication line is dependent upon several factors, namely, the number of bits which can be sent in a given time, the number of bits used to represent each character (frequently 8 bits) and the number of extra bits (that is start, stop and error parity bits) which need to be sent along with the data. The basic rate of data transmission is measured in bits per second (the *baud rate*). In practice, baud rates for computer communications are limited to values from 50 to 19,200 baud, with typical rates of 300, 1,200 and 9,600 baud. In a typical computer application, asynchronous data transmission may take place at 1,200 baud with 1 start bit, 8 data bits, 1 error parity bit and 2 stop bits. Since 12 bits are transferred at 1,200 baud the yield rate is 100 data units per second (1,200/12 = 100). The equivalent for synchronous transmission is 1,200/8 = 150 data units per second.

Software control

Software, as well as hardware, is required for computer communication. It is needed to establish and monitor connections (for example, sending/receiving data to/from a communication interface) and for changing communication parameters (for example, the baud rate, error parity and number of start and stop bits). The operating systems of the majority of computers have the appropriate commands necessary to communicate with peripheral devices (such as disk drives, printers and screens), but less common devices (such as digitizers, light pens and plotters) require additional software. The software commands for controlling communication operations, unfortunately, vary from computer to computer and from device to device, and so it is difficult to offer generalizations about them. The detail of how to communicate between specific machines may be obtained from manufacturer's manuals.

A number of packages exist that assist with data transmission between computers, but many of these use non-standard communication protocols and are machine- or operating-system specific (they only allow file transfer between similar types of computers or operating systems). One exception to this, which is becoming increasingly widely used, is the Kermit file transfer package designed by the University of Columbia, USA (Lee and Lee 1986). Kermit allows data to be transferred between any two computers which are running the package. Versions of Kermit are available free of charge for all common types of mainframes, minicomputers and microcomputers.

COMMUNICATING WITH OTHER COMPUTERS

The transmission of data between computers may take place via direct lines, local area or wide area networks, or telephone links (using telecommunication). These will be discussed in turn.

Direct lines

In situations where data transfers are required between two computers which are in close proximity (for example, in the same room), the simplest and cheapest method of communication may be the use of a direct line. These fairly unsophisticated communication links, connect the communication interfaces of two computers with a length of cable and employ a suitable file transfer package such as Kermit. The exact details of the transfer depend on the types of computer and software package involved in the transfer. Generally speaking, transfers are most easily arranged between similar types of computer with similar operating systems that are in close proximity.

Local and wide area networks

There is no formal agreement about what constitutes a *Local Area Network (LAN)*. As a rule of thumb, they may be described as digital communication facilities which carry data at high speed (0.1 to 100 megabits per second) over a limited range (0.1 to 10 km). In practice, this usually means within a single building, department or campus. A network with a range in excess of about 10 km is referred to as a *Wide Area Network (WAN)*. These usually make use of telephone links and, therefore, tend to be less clearly defined than LANs.

The components of a simple network are shown diagrammatically in Fig. 9.7. Any type of computer with an appropriate communication interface and software may be included in the network. The servers are combinations of hardware and software which facilitate the sharing of peripherals. The file server controls access to disks (hard or floppy). The printer server spools (queues) output to a shared printer. Disks, printers and other peripherals, such as digitizers and plotters, may still be linked directly to individual computers but in such cases cannot be used by other computers on the network. The gateway is a link to other networks (LANs or WANs) such as JANET (Joint Academic NETwork; see below). The connection media currently used in computer networks are virtually all bounded (that is wires and cables), although in some circumstances unbounded (that is, radio, microwave and infrared) broadcasts may be used. The principal types of bounded media are twisted-pair wire and coaxial cable. Twisted-pair wire is the plastic-coated twisted copper wire used for local telephone transmission. It is relatively cheap and can be easily installed. Coaxial cable offers the ability to send more than one signal at the same time (each signal is transmitted at a different frequency), faster transmission rates, high immunity to electrical interference and a low incidence of errors.

Networks can be classified in a number of ways, of which typology is

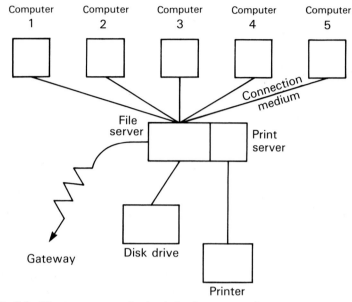

Fig. 9.7. The components of a simple local area network.

the simplest and most widely used. Three common types of network typology are shown schematically in Fig. 9.8. The diagram of the simple network already discussed (Fig. 9.7) is of the star type, as is the JANET wide area network described below. This network typology is commonly used in simple microcomputer networks. The best example of a ring network is the Cambridge ring developed at the University of Cambridge. The most widely used network typology is the bus network. Ethernet from Xerox, which has become the *de facto* standard for computer systems, and Acorn's Econet microcomputer network (see below) are examples of bus networks.

The connection of computers to form an integrated network system offers a number of advantages. Perhaps the greatest advantage is that it allows the sharing of expensive peripherals such as disk drives, printers and plotters. This is particularly effective for peripherals such as plotters which are used relatively infrequently. A second advantage is that it allows the sharing of software such as spreadsheets and statistical analysis packages and data files. A lecturer may, for example, create one copy of a data file of, say, fifty variables which students can access in order to test some particular geographical hypothesis. Third, a network provides the opportunity to establish an electronic mailing system, to aid information flow around and between departments and institutions.

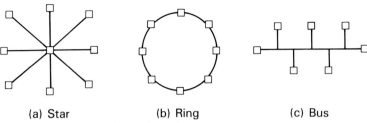

(a) Star	(b) Ring	(c) Bus

Fig. 9.8. Three types of network typology.

The first two of these facilities could be offered by a multi-user computer system. However, a centralized computer system has the disadvantage that overall control of the computer rests not with the user but, normally, with a computer manager. This makes the task of workload scheduling more difficult.

The Econet local area network

Econet is one of a number of local area networks which can be used to interconnect Acorn microcomputers and associated peripherals. It can be used to connect together up to 254 microcomputers and peripherals at distances up to about 1 km. The layout of a typical Econet system which might be used in a geography department is shown in Fig. 9.9. The central component of the system is the file server. This is a standard Acorn microcomputer fitted with a second processor and disk drive. When in use as a file server it may not be used as a microcomputer in the normal way. The remaining six microcomputers can be used to process data. To enable them to use Econet they all need special software. This controls data transfer between the microcomputer and the file server. In addition to the data processing facility, one of the microcomputers also acts as a print server. For this purpose it has additional software. The print server controls the spooling and printing of all jobs submitted for printing by users. The microcomputers are connected via a five-wire cable. One pair of twisted wires carry the data signal. The other pair carry the clock signal required for synchronous serial transmission of data (see the earlier part of this chapter). The fifth wire acts as a signal ground. The cable is fitted with a five-pin DIN plug which connects to the Econet socket at the rear of the microcomputer. Terminators, at the two ends of the cable, are required to prevent signal reflection and interference during transmission.

An Econet system is normally run on two levels. The highest level of knowledge and system access is required by the system manager who is responsible for setting up and running the system (for example changing and backing up disks). A second lower level of knowledge and access is

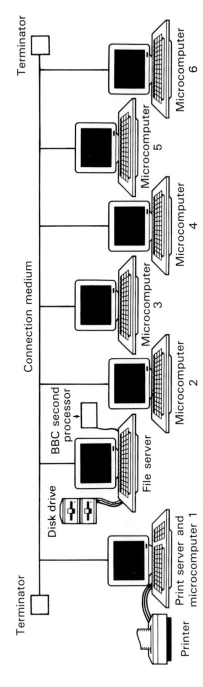

Fig. 9.9. The layout of a typical Econet local area network.

required by system users. In general, all the facilities available on the standard microcomputers are available under Econet (though some software may require modification). Apart from the obvious difference of being able to share peripherals and software, Econet provides a number of other useful facilities. To protect users' files, a security system of user-identification codes and passwords is established by the system manager. When users wish to gain access to the network they must first log on, by typing their individual user ID and password. Econet also provides a number of other specialist commands such as:

*NOTIFY which can be used to send a message to another user or microcomputer on the network;

*REMOTE which can be used to take control of another microcomputer on the network. For example, a lecturer may wish to take control of a student's screen in order to demonstrate a particular technique;

*VIEW which is used to display a copy of another microcomputer screen.

The JANET wide area network

JANET (Joint Academic NETwork) was established by the Computer Board, the Science and Engineering Research Council (SERC) and the Natural Environment Research Council (NERC), in 1984 (Wells 1986). The specific intention was to provide a single wide area network, using mutually agreed standards, to link together British higher education establishments, research laboratories and institutions. Many other countries have such networks, for example, a similar function is performed in Europe by Euronet and in North America by Bitnet. As far as end users are concerned JANET is tariff free. The network uses leased lines on British Telecom's national telephone network. There is also a gateway which gives access to British Telecom's own Packet Switch Stream (PSS) network and thence the International PSS. To gain access to the network, institutions require a leased telephone line, a modem and a computer with an appropriate communication interface. Users require a terminal linked to this computer and an identification code.

The network currently supports four main activities, namely, file transfer, electronic mail, job transfer to remote hosts and a software catalogue. It has already proved to be a great fillip to geographical research. It enables members of the research community to stay in contact, using the electronic mail facility, and to share and exchange programs and data using the file transfer facility. Of greater importance, however, are the opportunities which JANET offers for centralized data bases and computer processing. Even now there are a number of large data bases maintained at various institutions, that can be accessed using JANET. Examples include the Economic and Social Research Council Data Archive at the University of Essex (which keeps data sets on many geographical subjects), the Population Census data and digitized administrative area boundary files at the Universities of Bath, Edinburgh, London and Manchester and the NOMIS data base (see Chapter 10) at the University of Durham. Many of these and other sites offer access to large powerful computers and expensive software. The Cray supercomputer and the ARC/INFO geographical information system software (see Chapter 10), at the University of London, are good examples here. This trend towards centralized data bases and computer processing is a very significant computer development. It has already greatly increased the facilities offered to individuals and has substantially improved the cost effectiveness of computing.

One example of the use of JANET by geographers involves the geographical information system tutor, ARCDEMO, established on the VAX minicomputer at Birkbeck College, University of London (Green and Rhind 1986; see also Chapter 10). This on-line demonstrator can be accessed from any remote site which is connected to JANET. To use the demonstrator, local software, which controls access to JANET must first be run. The user's identification code and the address of the Birkbeck College VAX computer are then entered. Once connected to the Vax computer, a single command initiates the operation of the demonstrator. The whole process of gaining access to the demonstrator normally takes less than a minute. The tutor uses text and graphics screens to provide an introduction to the basic principles and applications of geographical information systems in general and the ARC/INFO system in particular.

Telephone links

Communication between computers, which are some distance apart, frequently makes use of the telephone network (Fig. 9.10). The telephone network is designed to transmit voices, which are analogue signals. Since computers usually output digital signals, it is necessary to connect to the telephone network via an analogue-to-digital/digital-to-analogue converter called a *modem* (which stands for MOdulator/DEModulator).

Basically there are two types of modem. A direct connect (also called a hardwired) modem is one which plugs straight into a telephone socket (in Britain the standard wall-mounted British Telecom jack socket) and a computer communication interface socket. Acoustic coupler modems are also connected to a computer communication interface socket, but the other lead is connected to the telephone handset. The modem has

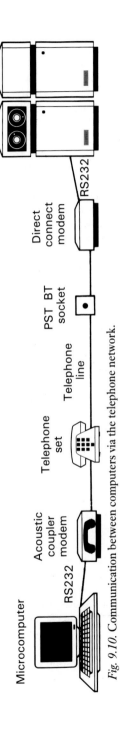

Fig. 9.10. Communication between computers via the telephone network.

163

two rubber cups into which a normal hand set fits. A speaker in the acoustic coupler modem plays the tones which are picked up by the microphone in the mouthpiece of the handset and transmitted down the telephone line. At the other connection a microphone in the acoustic coupler modem receives transmissions from the ear piece of the handset. Acoustic coupler modems have the advantage that they can be used wherever there is a suitable handset. On the other hand, they are more expensive and potentially more prone to interference. In Britain the increase in frequency of telephone jack sockets and unusual hand sets favours hardwired modems.

Modems come in a wide variety of shapes, sizes and makes. One of the most important features is the range of speeds for communication. The main standards for transmission/reception are 300/300 baud and 1,200/75 baud. Cheap modems allow only one transmit and one receive speed, whereas more expensive models offer a choice of several. The most common form of data transmission using telephones takes place simultaneously in both directions (full duplex). As this is relatively expensive, modems which allow non-simultaneous transmission in both directions (half-duplex) and transmission in only one direction (simplex) have been produced. Both mains and battery powered modems are available. Although the battery powered units are more expensive, this is compensated by the added freedom and flexibility. More sophisticated modems have facilities for storing telephone numbers (which can be selected at the push of a button), automatic redialling/answering and hardware and data error checking procedures.

Telephone links are available on either a leased line or dial-up line basis. Leased lines (also called private or dedicated lines) are contracted on a semi-permanent basis and can be used 24 hours a day if needed. They are usually used in situations where large quantities of data are frequently transmitted. Dial-up lines (also called switched lines) make use of normal telephone facilities. Calls are maintained as long as they are required and are terminated by replacing the hand set.

In order to reduce the cost of using the telephone network for data transmission in Britain, BT have developed a special national and international network called Packet Switch Stream (PSS). Users of PSS access the network by making a local call to the nearest network node. Access is restricted by the use of personal identification codes and passwords. In addition to being cheaper, PSS is much faster and more reliable. The NOMIS system described in Chapter 10 is one example of a data base which may be used via PSS.

VIDEOTEX

Videotex is a generic term for the data communication systems which emerged in the early 1970s and are now very popular. Viewdata and teletext are both types of videotex system. *Viewdata* ystems use modified computers and other terminals to receive and interact with text and graphics, which are held on other computers and are transmitted over the telephone network. *Teletext* systems use modified terminals, in most cases television sets, to receive (but not interact with) text and graphics held on other computers. Both teletext and viewdata systems offer users access to large data bases containing vast amounts of information on a variety of different subjects. In addition, viewdata systems provide a range of interactive facilities including holiday booking and home banking, the ability to download software and electronic mail for sending/receiving messages.

Ceefax and Oracle are the teletext systems available in Britain. Between them they have currently over 3 million home subscribers. They offer information about subjects such as rail and air travel, weather and up-to-date news. Charges are normally made on a quarterly rental basis. France and Canada are examples of other countries which also have teletext systems called, respectively, Antiope and Telidon.

There are many viewdata systems. Some are run by professional organizations and are often expensive to use, whilst others are run by amateurs and are free (beyond the cost of the telephone call). The most widely known professional public viewdata system in Britain is British Telecom's Prestel (see below). British Telecom also offer Telecom Gold which is mainly an electronic mail system for business users, with some public access data base facilities. Public viewdata systems are available in several other countries, for example France (called Teletel), West Germany (Bildschirmtext) and Japan (CAPTAIN). Other large professional systems, which may contain information of interest to geographers include The Source, Knowledge Index and World Reporter. Knowledge Index, (an American data base) for example, has twenty-seven subsections including data on biological abstracts, government publications, computers and education. Individual topics may be located using its comprehensive search facilities. Amateur viewdata systems are usually called 'bulletin boards' (derived from the facility to leave messages or bulletins in mailboxes). In spite of their name, many are well organized and have capabilities for file transfer and some have specialist information data bases. Guides to bulletin boards are frequently published in popular computing magazines (see for example *Personal Computer World* June 1987).

The recent upsurge in interest in viewdata systems in educational circles has resulted in the establishment of a number of special education systems. In Britain, the most well known are probably Micronet 800 and the Prestel Educational Service (provided by the Council for Educational Technology), which are a part of Prestel, and The Times Network for Schools (TTNS). The latter is dedicated exclusively to education and is sponsored by industry. It includes a central data base and an electronic mail network. At a local level several packages are available for setting up viewdata systems on microcomputers. Some can download data from other videotex systems as well as allow users to create their own pages.

Extracting information from the Prestel on-line data base

The Prestel viewdata system was launched commercially in Britain by British Telecom in September 1979. It comprises an electronic mail network and information data base of several hundred thousand pages. It can be accessed using special Prestel sets, ordinary TV sets and microcomputers fitted with Prestel adapters. Computer access requires suitable communication software and a modem which will transmit/ receive at either 1200/75 baud (the usual speeds) or 300/300 baud.

Prestel is different from most other viewdata systems in that it uses colour and graphics and is page-based. Most of the information on Prestel is supplied by information providers (IPs) such as the British Tourist Authority, Department of Employment, Economic and Social Research Council, Meteorological Office and Ministry of Agriculture, Fisheries and Food. IPs are responsible for deciding on the content and presentation of individual pages. British Telecom is responsible for maintaining the system and providing the indexes. The charge for using Prestel is made up of four parts, a quarterly standing charge, the cost of the telephone call to the Prestel computers, a charge for the time spent connected and a charge for viewing any pages for which the IP has fixed a price. These rates vary for different users, in different locations and at different times. There are over 50,000 Prestel subscribers, who annually make over 220 million frame accesses and mail over 4 million messages. Within Prestel there are a number of private pages established by Closed User Groups (CUGs). Examples include Micronet 800, the Prestel Educational Service 880 and Viewfax 258, which have considerable educational following. Access to CUGs is restricted to users who pay an extra subscription or are given access by the information provider.

The Prestel data base has a number of entries which are of interest to geographers. A selection of items from the index is shown in Fig. 9.11. It includes economic statistics, education, EEC, employment, environment and Egypt. Other items of particular interest might be the popula-

Aberdeen	Netherlands
Adult education	
Aerial photography	
Air fares	On-line information services
	Parliament
Barnsley	Polytechnics
Books	Population statistics
	Prestel
Cities	
College of Higher Education	Qatar
Computers	
Current Affairs	
	Russia
Degrees	
Devon	Schools
	Science
	Statistics
Economic statistics	
Education	
EEC	Tourism
Employment statistics	Transport
Environment	
	United States of America
	Universities
Fiji	
Government publications	Videotex
Government statistics	
	Weather
Hong Kong	Word processors
Information Technology	Yemen Arab Republic
Inns and pubs	
Isle of Man	
	Zaire
Japan	
Kenya	
Libraries	
Mailbox	
Microcomputing	

Fig. 9.11. A selection of 'geographical' topics available in Prestel.

tion data (provided by the Office of Population and Census Survey), weather reports and forecasts (provided by the Meteorological Office) and geographical profiles of countries.

A Prestel session is initiated (assuming that the Prestel set is plugged in and turned on) by dialling the appropriate British Telecom telephone number and connecting the modem when the number is answered. The user's personal identification code and password are then entered into the Prestel terminal and Prestel will answer by displaying the welcome page. There are several ways of finding information in Prestel. There is an alphabetically arranged on-line subject index and a facility to browse through a series of menu pages and then choose a subject to examine in more detail. Once a particular subject or page is located it can be accessed by simply entering the page number. To assist users, Prestel maintains a number of pages of information about the system, for example, how to use the system, details about the bill to date and new topics on the system. Prestel, like all other viewdata systems, offers some interactive facilities. This may take the form of typing information, such as names and addresses, into response frames in order to obtain brochures or goods and services. Prestel also offers an electronic mail service. Some examples of Prestel pages are shown in Fig. 9.12.

Prestel offers up-to-date information, in a simple language that incorporates many graphic images and maps. It has a versatile hierarchical data base structure and page format, which can easily be adapted to fit the requirements of many information providers and private users.

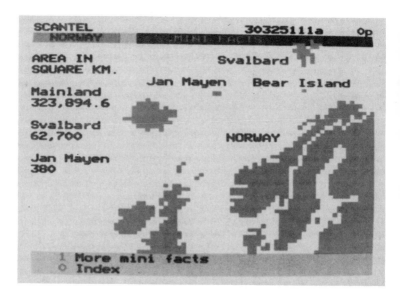

```
Government Information   5004113a      0p
   OPCS
POPULATION: regional       (thousands)

Region              mid-1971         mid-1984

England             46,412           46,956
Wales                2,740            2,807

North                3,152            3,093
Yorkshire &
  Humberside         4,902            4,904
East Midlands        3,652            3,874

East Anglia          1,688            1,940
South East          17,125           17,112
 (Gtr London         7,529            6,756)

South West           4,112            4,461
West Midlands        5,146            5,176
North West           6,634            6,396
   Key 9 Population index    0 Main index
```

Fig. 9.12. Two example pages from Prestel.

Compared to other media it is a relatively cheap way of obtaining information (someone has to pay for library books). On the other hand, only a limited amount of information is of geographical relevance and many maps are of poor quality. Because Prestel is used for many purposes, on many types of equipment, it does not make full use of the hardware and software potential offered by some computers. There are, for example, only limited search facilities. It is also difficult for inexperienced users to find their way around the data base. Furthermore, as the costs are highest during teaching hours and there are copyright problems related to copying pages to other systems, many institutions have not made full use of its facilities.

CONCLUSION

This chapter has considered the basic principles of computer communication and the main types of computer link. Details have also been presented about Videotex, a computer application predicated upon good computer communication, and an example of how to access the Prestel data base. The nature of the subject matter has made this chapter necessarily more technical than other chapters. This is justified in view of the importance of computer communication which plays an important role in many computer applications. In the future as the trend towards large centralized data bases with distributed computer access increases, computer communication will become an increasingly important topic.

FURTHER READING

Gee, K. C. E. (1982) *Local Area Networks*. National Computer Centre Publications. (A good readable book on LANs which includes several examples and applications.)

Megarry, J. (1985) *Inside Information: Computers, Communications and People*. British Broadcasting Corporation. (Very good descriptions of how telephones and telecommunications work. It also has sections on videotex.)

Seyer, M. D. (1984) *RS232 Made Easy: Connecting Computers, Printers, Terminals and Modems*. Prentice-Hall, Englewood Cliffs, New Jersey. (A very good, easy to understand discussion about the principles of communication. Also contains a wiring guide for many types of devices.)

Zorkoczy, P. (1985) *Information Technology: an Introduction* 2nd edn. Pitman. (A well produced text which deals with the basic principles and applications of communications.)

10

Geographical information systems

Preceding chapters in this book have examined key facets of geographical computing and the fundamental concepts relevant to computer use in geography. Examples have been presented describing the use of computers for a range of geographical applications, including the collection of soil profile data, the analysis of rainfall data and satellite imagery and the mapping of population statistics. Although it has been necessary, for obvious practical reasons, to consider each type of application in a separate chapter, this convenient organization device is somewhat artificial. Even so, it does reflect the way in which geographers have, until relatively recently, viewed geographical analysis and computing. Indeed, it is only in the past few years that geographers have started to appreciate the real benefits to be derived from integrating data and programs into large comprehensive systems, thereby opening up the prospects of many different operations and potential applications.

Such integrated systems for the collection, storage, manipulation and presentation of geographical data, are referred to as *Geographical Information Systems (GISs)*. Functional GISs comprise an integrated collection of computer hardware, computer software and geographical data. They exist in a variety of forms and embody the potential for an enormous range of applications. Those which have been established for specific applications are sometimes given other names such as Land Information Systems (LISs), Natural Resource Information Systems (NRISs), Soil Information Systems (SISs) etc., but they are all developed according to the same basic GIS principle.

This chapter describes the fundamental characteristics and key features of GISs. The benefits and functional components of GISs are discussed and a number of examples of their application are presented.

THE BENEFITS OF GISs

The significance of developments in geographical information systems has been recognized by many individuals and organizations. The British Government, for example, commissioned a special committee of enquiry to report on handling geographical information (DoE 1987). According to this report, GISs offer quick and easy access to large volumes of data. The key to their enormous value is that they offer the user the opportunity to analyse and manipulate large data sets – to select data by area or theme, to search for particular characteristics or features in areas, to update data sets quickly and cheaply, and to model data and assess alternatives. They can also be used to create new and varied types of output in the form of maps, graphs, address lists and summary statistics – all tailored to individual needs.

GISs are in the process of transforming traditional methods of geographical analysis. Already they present geographers with a greatly enhanced opportunity to describe, explain and predict spatial patterns and processes. They provide the ability, using large data sets, to build more sophisticated and realistic models and to test more probing hypotheses than has hitherto been possible. For example, a GIS containing biogeographical data for northern Africa, will allow point-based plant quadrat data to be linked with area-based soil data and line-based animal migration flows. Using a system like this it would be possible to ask such questions as how much land with a certain kind of vegetation cover and a certain soil type has more than a particular number of animals walking across it during a year? Clearly such information could be crucial for describing, explaining and predicting erosion rates in some semi-arid areas.

The benefits of GISs to date and, more importantly, the potential benefits they offer for the future, cannot be overemphasized. The enormous recent advances in geographical information systems must be seen as more than just technological and methodological change. Whilst they do offer geographers new tools which will significantly help to improve our understanding of spatial patterns and processes, the use of GISs may stimulate the development of a new philosophy which integrates the work of human and physical geographers, climatologists, hydrologists, soil scientists, planners, transport geographers and the like.

Finally, by their very nature GISs are application orientated. They offer geographers the chance to become more technologically aligned and to increase their perceived relevance to the world at large, by developing a more problem orientated approach to the subject. This point is borne out by the examples which are discussed below.

THE DEVELOPMENT OF GISs

A demand for geographical information systems has been created by four main factors. First, the great proliferation of data about the environment in recent decades and, more especially, the vast increase in the quantity of data available in computer format has exercised a crucial impact. For example, many national mapping agencies, such as the United States Geological Survey (USGS) and the Ordnance Survey of Great Britain (OS), have embarked on major projects to create country-wide digital map data bases at a variety of scales. Second, recent advances in geographical theory and techniques, which have outgrown the capabilities of existing computer systems, have also led to a need for sophisticated and integrated computer systems. The multi-dimensional nature of geographical data is a third important factor. Conventional computer data base management systems were designed to handle one-dimensional data and cannot cope satisfactorily with two- and three-dimensional geographical data, much of which are available on different spatial bases. There was a need therefore to develop new systems capable of handling large quantities of geographical data. The final factor stems from the practical nature of GISs. Members of commercial and government agencies quickly realized the enormous commercial value of GISs for a wide range of applications, such as locating mineral resources, managing property registers and market analysis. A consequence of this is that much of the research and development work on GISs has been undertaken by commercial organizations and is strongly application-orientated.

The principle of the GIS was first conceived in the early 1960s and the first system, the Canadian Geographic Information System (CGIS) was implemented in 1964 (Smith *et al.* 1987). However, despite some progress in the late 1960s and 1970s, it was not until the 1980s that they really came into widespread use. Until this time major technological limitations restricted their advancement. However, the rapid reduction in computer hardware costs during the 1960s and 1970s meant that, by the beginning of the 1980s, hardware was no longer such a limiting factor. Significant in reducing the cost of hardware has been the enormous developments in microprocessors. Accompanying these hardware developments, there have also been great increases in both the quality and quantity of commercially available packaged software. This has both reduced the development costs and has made GISs easier to use. The advances in computer operating systems and human-computer interfaces have been significant here.

THE FUNDAMENTAL COMPONENTS OF GISs

GISs are normally classified into two generic groups depending on the way in which they handle geographical data. The first type, *raster-based* GISs, deal with data encoded in grid cell format. Many of the GISs which utilize significant quantities of remote sensing data tend to favour this type of format (Curran 1985). The second type, *vector-based* GISs, deal with data encoded as vectors using cartesian co-ordinates. This type is favoured by users of topographic and thematic map data. The relative merits of these two systems have already been discussed in Chapter 5.

The most important part of any GIS is the data. This is not least because geographers often deal with exceptionally large quantities of data. In a GIS, geographical locational data – such as administrative boundaries, point locations of sites and grid squares – are used to provide a spatial reference for statistical (also called non-locational or attribute) data, such as census counts, measurements of stream levels and satellite sensor readings.

The actual configuration of computer hardware and software used to process and store data depends upon the specific applications of a GIS. All GISs require hardware for the basic operations of data input, storage, manipulation and output. The particulars of the hardware necessary to perform these operations is described in other chapters (see especially Chapters 5 and 11). Generally speaking, because of the very large quantities of data, it is important that the computer hardware is equipped with a powerful processor and a large storage capacity.

Presently, a great variety of software is available which is capable of performing the standard operations required by a GIS. Indeed, there are in excess of a hundred packages worldwide which purport to be GISs. Although no comprehensive catalogue has yet been compiled, a useful review of sixty-one microcomputer-based GIS software packages is provided by Gray and Maizel (1985). The most successful GIS packages consist of two major software elements. A data base management system (DBMS) is used to handle the basic data storage, management and analysis operations. Alongside this there is a graphics package to handle the equivalent operations on geographical data. For example, the ARC/INFO GIS (probably the market leader at present) is one such hybrid system comprising ARC, a graphics package, linked to INFO, a relational DBMS. It is available for both minicomputers and microcomputers and has been used for many applications (see below for further details).

THE FUNCTIONS OF GISs

Geographical information systems are required to perform a great many different functions. Unfortunately, the way these functions are implemented is conditioned by the nature of the hardware and software which form the GIS. The aim here is to present a generic introduction to some of the most commonly used GIS commands. A wider, more technical discussion may be sought in Dangermond (1983) and Burrough (1986). GIS functions may be divided into four basic groups: input, storage, manipulation and output (Fig. 10.1). Because GISs are

INPUT	Data collection – both geographical and statistical
	Data transfer
	Data verification
	Data editing
STORAGE	Disk (temporary)
	Magnetic tape (more permanent)
MANIPULATION	Cartographic functions – scale changes
	– vector-raster-vector conversion
	– projection changes
	– map embellishment (add scale, title, north point and legend)
	Data integration – map overlay
	– spatial aggregation
	– spatial transformation
	Feature measurement – number of features
	– calculate distance, area, volume and shape indices
	Spatial searching – on points, lines and areas to determine distance, angle, overlap and inside
	Statistical analysis – descriptive statistics, cross-tabulation, correlation
OUTPUT	Data presentation – maps
	graphs
	tables
	text
	Data transfer

Fig. 10.1. Some typical functions of geographical information systems.

application-orientated, it seems appropriate to describe the hardware, software and data requirements of a functional GIS in the context of an example application. The GIS proposed for health care planning, outlined by Maguire and Mohan (1985), will serve this purpose.

Health care planners have long been hampered by the technical problems encountered in linking together population census and health service activity data, which are only available on different spatial bases. Whilst census data are available for administrative areas (enumeration districts, wards, districts etc.), health service activity data are only available for postcode areas (patients are allocated to areas on the basis of their home address), point locations (hospitals, general practitioner surgeries etc.) and line flows (patient mobility data are available as migration flows along transport routes). The GIS designed by Maguire and Mohan (1985), was intended to assist health care planners by allowing the integration, analysis and mapping of these data. Such a system could assist in the monitoring of service delivery and the identification of target areas for future action and research.

Data input

Both geographical and statistical data need to be input to a GIS. The process of data input usually involves data collection, verification (checking) and editing (see Chapter 2 for discussion of the theory and practice of these operations). A digitizer is usually the only additional peripheral device which is needed to collect these data, although occasionally data loggers and other specialist equipment can be employed. In some instances the required data are already available in computer compatible format and so it is desirable to have a facility for their input into a GIS on magnetic tape, floppy disk or via an electronic network. In the case of the GIS for health care planning (Fig. 10.2), the geographical data are the digitized boundaries of areal units, lines and points. The statistical data are from the census statistics, health service activity records and transport surveys.

Storage

There is an obvious requirement in GISs for temporary and more permanent storage of large quantities of data. Floppy disks are normally used for the former and magnetic tapes are used for the latter. GISs normally utilize standard computer operating system commands for performing basic storage and retrieval operations. Any one of a number of different formats may be chosen for storing the data, depending both on the type of data and the application (see Chapter 3 for further

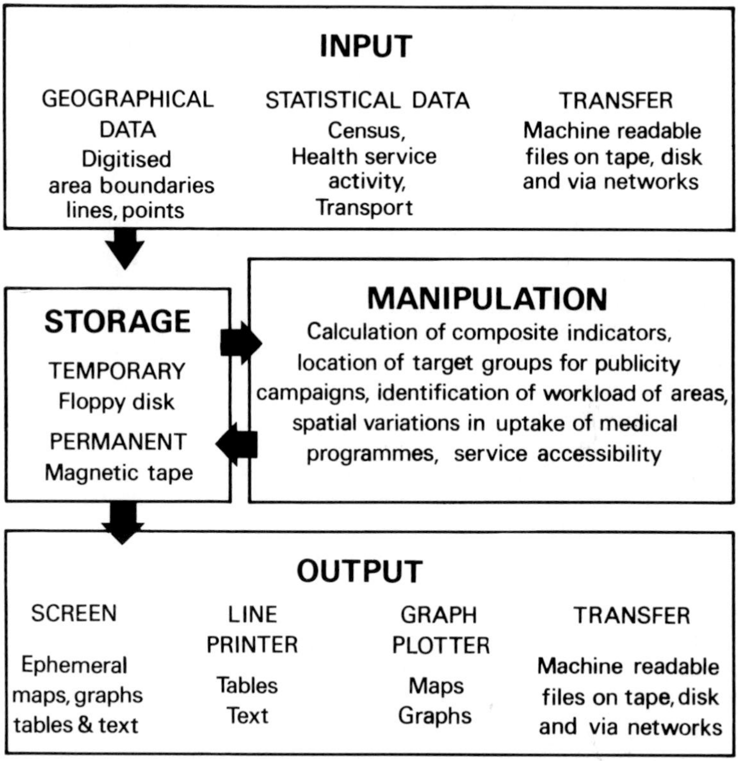

Fig. 10.2. A geographical information system for use in health care planning.

discussion of this topic). In the case of large GISs, data storage may be one of the largest costs.

Manipulation

The manipulation of data is a fundamental aspect of any GIS. Depending on the application, GISs may be required to perform a great variety of functions and a selection of the most commonly used is shown in Fig. 10.1.

Cartographic functions are often the first type of manipulations to be performed when a GIS is put to work. Cartographic manipulation may involve changing the map scale, converting the data from raster to vector or vector to raster format, changing the map projection, or the embellishment of maps (by the addition of a title, scale, north point and legend).

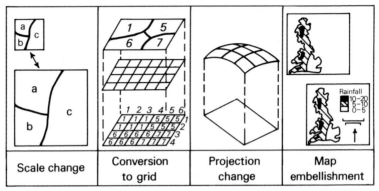

Fig. 10.3. Some cartographic functions of geographical information systems. *Source*: redrawn from Dangermond 1983.

Data integration is probably the function for which GISs are best known. Basically, this entails bringing together disparate data sets, which are available on different spatial bases, in order that meaningful analyses may be undertaken on them. Although simple in principle, data integration can in fact require a great deal of computation. It mainly involves overlaying different data sets (often called layers) and then performing arithmetical and relational operations on them. This might entail, for example, adding, subtracting, or multiplying the 'data layers'. The data integration function facilitates fundamental GIS analysis questions about intersection and coverage to be addressed. Returning to the example of the health care planning GIS the *intersection* question could be posed in the form: where are the areas that are greater than 1 mile from a hospital, in which more than 7 per cent of the households

Search criteria	Data layer	Spatial units	Data source
> 1 mi	hospital location	points	health authority
> 7%	one parent families	administrative areas	population census
> 5%	young children	administrative areas	population census
< 70%	vaccination	postcode areas	health authority
> 10 mins	bus route	bus network	city transport

Fig. 10.4. The search criteria and data characteristics used to illustrate the principle of data integration (see also Fig. 10.5).

are one parent families, in which more than 5 per cent of the households have young children (aged 0–4), in which less than 70 per cent of the children have been vaccinated against whooping cough and which are greater than 10 min walk from a bus route? A GIS set up with appropriate data layers (Fig. 10.4) could be searched using these criteria (Fig. 10.5), which might be crucial for locating groups in need of health care. For example, a 1 mi buffer would be placed around all hospitals and a 10 min walk buffer would be placed around bus routes. The relevant area-based data sets would also be searched to extract appropriate areas. These would then be combined by overlay to locate the relevant area of intersection. The *coverage* question is really just a variation of this process and may be posed in the form: what are the characteristics of the

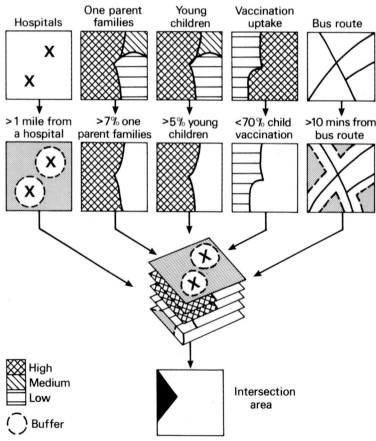

Fig. 10.5. The principle of data integration.

areas with high vaccination uptake levels? The information derived from highly refined analyses such as these could be important for identifying spatial variations in the uptake of medical programmes, for identifying the location of target groups for publicity campaigns and for assessing the accessibility and workload of services.

In some instances data integration will necessitate changing the spatial organization of one or more data layers so that the data are comparable with other layers. At its simplest this might mean aggregating one type of area to give another type or else selecting the point closest to an area for the purposes of comparison. However, considerably more processing may be involved in the integration of some data sets.

Feature measurement includes a number of operations which may be carried out on geographical entities in one or more data layers (Fig. 10.6). It involves operations such as counting the total number of occurrences of particular features (such as hospitals), measuring distances between objects (such as towns), calculating areas (such as field size), calculating volumes (such as the amount of material to be moved during road construction) and calculating shape indices (of such things as sand dunes).

Spatial searching is also a well known functional aspect of GISs. Basically, spatial searching entails locating certain features within a spatial data base using some type of spatial search criteria. The features in question may be points (hospitals, water holes, hypermarkets etc.), lines (bus routes, animal migration routes, rivers etc.) or areas (census tracts, hydrological catchments, television viewing areas etc.). The four basic search criteria that may be used are concerned with the distance between objects, the angle between objects, the overlap between objects and whether one object lies within another. The possible combinations of features and search criteria are shown in Fig. 10.7. Some combinations, of course, are not applicable; thus it is not appropriate, for

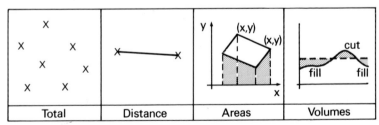

Fig. 10.6. Some feature measurement functions of geographical information systems.
Source: redrawn from Dangermond 1983.

		I	Point	I	Line	I	Area	I
	point	I	*	I	*	I	*	I
(a) Distance	line	I	*	I	*	I	*	I
	area	I	*	I	*	I	*	I
	point	I	*	I	*	I	*	I
(b) Angle	line	I	*	I	*	I	*	I
	area	I	*	I	*	I	*	I
	point	I	*	I	*	I	*	I
(c) Overlap	line	I	*	I	*	I	*	I
	area	I	*	I	*	I	*	I
	point	I	Not app.	I	Not app.	I	*	I
(d) Inside	line	I	Not app.	I	Not app.	I	*	I
	area	I	Not app.	I	Not app.	I	*	I

Fig. 10.7. Combinations of feature type and search criteria used in GIS spatial searching. Not app. = not applicable.

example, to try to determine whether one point is inside another. The four basic search questions which may be asked are:

1 What is the distance from feature X to feature Y?
2 Is feature X within the angle θ from feature Y?
3 Does feature X overlap with feature Y?
4 Is feature X inside feature Y?

The answers to these questions may involve one or more data layers. Multiple search criteria may also be built up using these basic questions and many variations on the general themes are possible, as illustrated for example, in the intersection problem using the health care GIS already discussed above.

A further example, of a GIS designed to examine the impact of sulphur dioxide air pollution on the plant life in a local authority planning area, usefully illustrates some of these points. The three data layers in the GIS contain data about vegetation cover, the location of pollution sources and local authority planning areas. The effect of the pollution may be assessed, therefore, by using the GIS to extract all plant quadrat records which have less than 5 per cent of lichen type A (a lichen type which will not grow in areas of high pollution and is thus a good indicator species), which are in the quadrant north-east of the sulphur dioxide pollution source (thought to be downwind of the prevailing wind direction), which are within 10 miles of the pollution source and which are inside the planning area.

GISs are also required to perform various *statistical analyses*. This may take the form of summarizing and describing data, producing statistical profiles, assessing relationships between feature distributions (using, for example, regression analysis), trend surface analysis and network analysis for routeing purposes.

Output

The output from GISs may take several forms, of which the principal types are maps, graphs, tables and text. These can be drawn quickly on a computer screen. However, since these are often hard to evaluate and appreciate (due to the limited resolution of computer screens) and because they are ephemeral, some form of hard-copy device will also usually be necessary. Text and tables could be output to a line printer and a graph plotter will probably be most appropriate for maps and graphs. As it will frequently be desirable to transfer data to other computer systems there is also an obvious need for output in the form of machine readable files.

THE APPLICATIONS OF GISs

The applications of GISs are legion. Already they are used for a vast range of projects in government, commercial and academic environments throughout many parts of the world. Tomlinson (1984) estimates that in 1983 in the USA alone there were about 1000 GISs in operation and that by the end of the decade the total will have increased fourfold. By way of introduction to the application of GISs, DoE (1987) present brief details of sixteen major areas in which GISs have been used. These include land use and natural resource management, utilities (electric, gas, telephone and water) network management, property registration and development, market analysis and business location, mineral exploitation, regional development policy and teaching. It is, therefore, not possible to attempt to summarize even the current applications of GISs here. Instead, it seems more appropriate to present four selected case studies which illustrate the ways in which GISs have been used.

Planning and managing public services using GISs

Local and central government bodies, in several countries throughout the world, are using GISs for organizing and integrating data in an attempt to improve the efficiency of service provision. A selection of the types of projects for which GISs are being used is shown in Fig. 10.8.

Monitoring changes	in resources (land and building, equipment and infrastructure) and conditions (economic, social, demographic, environmental).
Forecasting changes	in housing requirements, in school rolls, in travel patterns, in the economy and in the demand for land, leisure, and community services.
Service planning	through identifying and forecasting changes in patterns of need for services and investments as a basis for the delivery of services and deployment of resources. This will determine both the scale of provision and its location; it will also highlight areas of social deprivation.
Resource management	e.g. building maintenance, refuse collection, grass cutting, route scheduling of supplies vehicles, mobile libraries, social service ambulances.
Transport network management	including provision and maintenance of highways, public transport schedules, school transport, street cleaning.
Public protection and security systems	e.g. police command and control systems, definition of police beats, location of fire hydrants, patterns of crime and incidents of fire.
Property development and investment	including the preparation of development plans; assessing land potential and preparing property registers; promoting industrial development; rural resource management.
Education	use of a wide range of data for teaching purposes, including the use of demonstration software packages.

Fig. 10.8. Some government uses of geographical information systems.
Source: DoE 1987.

The National On-line Manpower Information System (NOMIS) is a good example of the application of GISs for planning and managing public services (Townsend *et al.* 1986; Townsend, Blakemore and Nelson 1987). This system, which is based at the University of Durham in the UK, stores government data on employment, unemployment, job vacancies and population characteristics. These data are available at a variety of spatial scales, for a variety of spatial units and in the form of raw counts, tables, graphs and maps. The system offers a number of analytical procedures including a wide range of summary and comparison statistics (such as location quotients, chi square and regression). NOMIS is designed to be used by novice computer users and can be

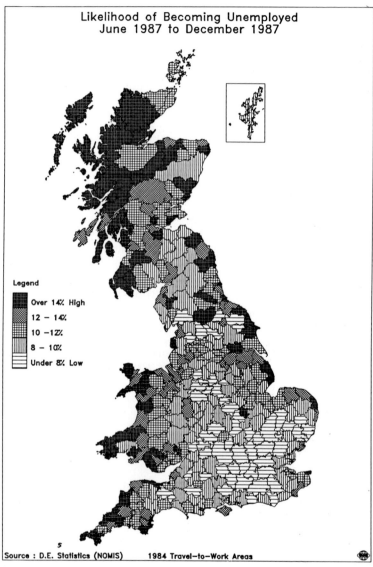

Fig. 10.9. An example of cartographic output from NOMIS showing the likelihood of becoming unemployed in 1987.

accessed using microcomputers and other types of computer terminal via the telephone network. Operational since 1982, NOMIS has been extensively used by government employees and academics for a number of purposes. Specific projects include the analysis of changes in employment and the rural–urban employment shift, population projections, the identification of areas of social deprivation and the examination of spatial variations in overcrowding. An example of NOMIS output is presented in Fig. 10.9.

Using GISs in natural resource management

North America has emerged as the world leader in GIS applications and natural resource management, especially forestry, has for some time been in the vanguard of developments (see Webb 1982). As early as 1982, thirty-two states in the USA possessed statewide natural resource information systems (Johannsen and Sanders 1982) and typical of these is the Minnesota Land Information System (MLIS).

The state of Minnesota established a land management information centre in the early 1970s to act as a data clearing house and information bureau (Minnesota Planning Information Centre 1987). It created a statewide 40 acre grid cell (raster) mapping system containing a range of natural resource data. Over the past ten years the system has been used for projects such as: an environmental assessment of the possible impact of copper-nickel mining; a study of seasonal homes and recreational activity; highway and powerline corridor routeing; solid, hazardous and radioactive waste disposal siting; and modelling the vulnerability of aquatic and terrestrial areas to acid precipitation. Since 1982 the Centre has been using the ARC/INFO GIS software to manage the system. Topographic data are entered from the United States Geological Survey (USGS) 7.5 minute (1:24 000) topographic maps. These are combined with thematic information about soils, geology, land use/land cover, public ownership, hydrology, transportation and political subdivisions, some of which are derived from remote sensing satellite data.

Two examples of the applications of the Minnesota Land Information System are shown in Figs 10.10 and 10.11. Figure 10.10 shows a statewide map giving details of census tracts identified by the Minnesota Pollution Control Agency (MPCA), on the basis of their social and economic characteristics, as potentially susceptible to the effects of soil lead pollution. Figure 10.11 is a map of 1986 election return data for the metropolitan area. It shows votes cast by members of the gubernatorial race for republican and democrat parties. Both of these types of data are held in the system because of their value for strategic planning.

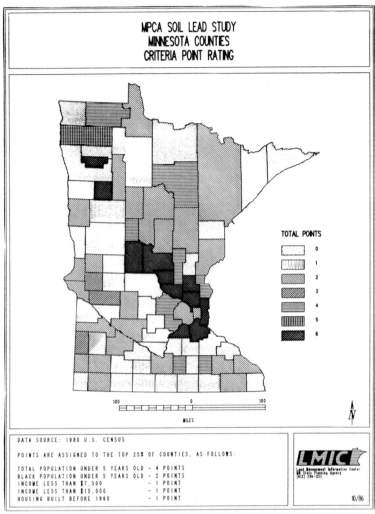

Fig. 10.10. Map of Minnesota state showing census tracts identified as
potentially susceptible to the effects of soil lead pollution. (Note
that the original was in colour.)
Source: provided by K. Pekarek.

Creating an environmental data base for the European Community

The task of managing and protecting the environment of the member
states of the European Community (EC) lies with the Directorate

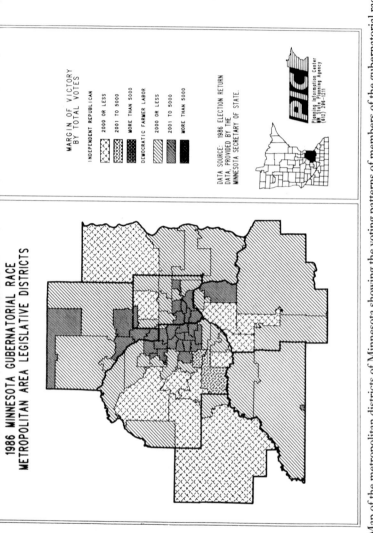

Fig. 10.11. Map of the metropolitan districts of Minnesota showing the voting patterns of members of the gubernatorial race during the 1986 election. (Note that the original was in colour.)

Source: provided by K. Pekarek.

187

General of the Environment, Consumer Protection and Nuclear Safety. To perform its functions properly this special directorate really requires an integrated spatial data base of environment descriptors. This requirement led to the initiation of a major project called CORINE (Co-ORdinated INformation on the Environment) which is funded by the EC (Rhind *et al.* 1986; Wiggins *et al.* 1987). The project encompasses three basic principles. First, since limited resources are available existing data must be utilized wherever possible. Second, there is an emphasis on collaboration between organizations of different member countries. Third, the data are needed at many spatial scales ranging from 1 km² to the size of countries. Initially, at least, the data base has been established using the ARC/INFO GIS and is based at Birkbeck College, University of London, in the UK. The data base contains information about topography, soils, climate, biotopes (sites of special biogeographical interest), birds, vegetation and less favourable agricultural areas. These have been collected from maps at scales of 1:3,000,000 to 1:500,000. This data base will eventually be expanded to

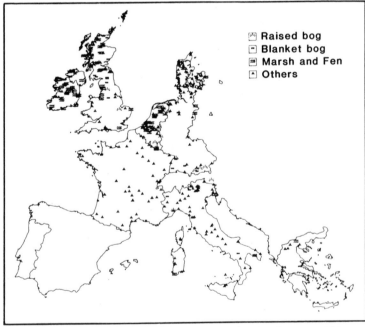

Fig. 10.12. Output from the CORINE data base showing the extent of the area covered and the distribution of biotopes.
Source: Wiggins *et al.* 1987.

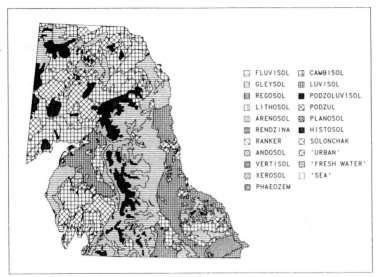

Fig. 10.13. Output from the CORINE data base showing the distribution of soils in northern England.
Source: Wiggins *et al.* 1987.

incorporate pollution, population and other data derived from paper maps, remote sensing data and statistical aggregates. Figures 10.12 and 10.13 show examples of output from the data base.

It is envisaged that the system will be used for a great many purposes. For example, it will be used to maintain a register of environmental information about the EC, to co-ordinate the biotopes inventory for the EC, to collect and organize data on water resources in the Mediterranean and other specific regions, to undertake a feasibility study into collating data on pollution (especially acid rain), and to identify areas threatened by erosion and, therefore, in need of protection.

The BBC interactive videodisk data archive

One of the most important functions of GISs is that they act as repositories of information; indeed they are very widely used to inventory resources. Any system required to store large quantities of data will clearly need to have an efficient and cheap storage facility. The so-called Domesday Project, which set out to mark the 900th anniversary of the original Domesday survey (see Chapter 1) by presenting a contemporary snapshot of Britain, represents an example of this kind of system. The Project was a collaborative venture between the British Broadcasting Corporation (BBC), the British Government and industry. It aimed to develop a national archive in the form of an interactive videodisk system combining digital data, text, photographs and sound, drawn from a wide variety of sources. Much of the data have a spatial component and so most of the uses will be manifestly geographical (Goddard and Armstrong 1986; Openshaw, Rhind and Goddard 1986). the Domesday System actually comprises a microcomputer with a high resolution screen, software, and a laservision videodisk player, which uses a laser beam to read two 12 in videodisks. The videodisks, called the Community and National disks, store vast quantities of information capable of being quickly retrieved, analysed and displayed using the computer software.

The Community disk contains a complete set of 1:50,000 Ordnance Survey topographic maps of Great Britain, together with text and pictures collected by schools, community groups and individuals in a special national survey carried out in 1986. Apart from the factual content, this disk also provides unique insights into the way the people in various areas of Britain view themselves. Areas and places can be located by searching the text and photographs, and the software also allows simple spatial analyses to be performed, such as calculating areas and distances.

The National disk represents a gigantic nationwide data base. It holds text essays, published articles, photosets from private and public collections, more maps and photographs and a enormous quantity of digital data. The data are divided into four categories: *culture* – arts, communication and tourism etc.; *economy* – industry and people at work etc.; *society* – education, health and population etc.; and *environment* – agriculture, climate, soils and water resources etc. Information on these topics is available for a wide range of different spatial units and different levels of spatial aggregation, using grid squares as the building blocks. Users can control the class intervals, colours and levels of aggregation. The data are accessed using a hierarchy of maps, a thesaurus of keywords, and a gazetteer of place names. Simulated walks (coordinated sets of photographs) and short films with sound tracks are also available. The statistical information on the disk can be presented in a variety of configurations including maps, bar charts and line graphs. Computer graphics can be overlaid on video map images and data can even be down-loaded to floppy disks for further analysis.

The system has many potential uses in government departments, libraries, commerce, industry and education. For example, estate agents could use it to provide information about the location of houses, retail companies could use it to obtain information about potential markets and store locations, rural and urban planners could use it to assess the level of service provision and journalists might use it to locate informa-

Fig. 10.14. Output from the Domesday Interactive Videodisk System.

tion about people, events and places. Perhaps its greatest potential is in teaching where it can be used to illustrate the concepts of spatial aggregation, the modifiable area unit problem and the technique of area measurement amongst other things. Examples of output from the system are shown in Fig. 10.14 and Plate 4.

CONCLUSION

This chapter has presented details about the basic characteristics of geographical information systems. The benefits and functional components of GISs have been discussed and some examples of major applications have been described. GIS is a tremendously exciting area of geography to be involved in at the present. The rate of technological development and the speed at which research is progressing ensure that new ideas and techniques are being presented almost daily. Unfortunately, this aspect of geographical computing like all others does face some significant problems relating to the availability of liveware and data. It is possible that this area of computing may take on even more significance in the future. The integrative and applied nature of GISs makes them very attractive to many geographers. They may help to bring together the many disparate branches of the discipline and may provide a common focus perhaps once again based upon the region as the fundamental geographical concept.

FURTHER READING

Burrough, P. A. (1986) *Principles of Geographical Information Systems for Land Resources Assessment. Monographs on soil and resources survey* **12**. Clarendon Press. (A little technical in parts but the best review of the principles of GISs.)

DoE (1987) *Handling Geographic Information*. HMSO. (An excellent introduction to the applications of GISs.)

Dangermond, J. (1983) A classification of software components commonly used in geographic information systems. In Peuquet, D., O'Callaghan, J. (eds) *Design and Implementation of Computer-based Geographic Information Systems*. IGU Commission on Geographical Data Sensing and Processing, Amherst NY. (A readable introduction to GIS software functions.)

Jensen, J. R. (1986) *Introductory Digital Image Processing: a Remote Sensing Perspective*. Prentice-Hall, Englewood Cliffs New Jersey. (Chapter 10 on 'The interface of remote sensing and geographic information systems' is a good up-to-date review of GISs from a remote sensing perspective.)

Marble, D. F., Calkins, H. W., Peuquet, D. J. (1984) (eds) *Basic Readings in Geographic Information Systems*. SPAD Systems, Williamsville New York. (A useful collection of essays on various aspects of GIS.)

Smith, T. R., Menon, S., Star, J. L., Estes, J. E. (1987) Requirements and principles for the implementation and construction of large-scale geographic information systems. *International Journal of Geographical Information Systems* 1: 13–31. (A modern introduction to the principles of GISs.)

11

Computer hardware

This chapter is concerned with the fundamentals of computer hardware. The structure and development of computers are outlined and there is consideration of the basic elements of computer hardware, including discussion of processors, input, output and storage devices, and peripherals. It is not envisaged that all the material in this chapter will be of primary interest to all geographers. Rather it is seen as a chapter for those interested in learning more about the potential limitations and suitability of computers for geographical work and for those interested in developing their own applications.

Computers are programmable electronic machines which can input, manipulate, store and output data. They come in a variety of shapes, sizes and makes. Broadly, there are three main types which are called digital, analogue and hybrid computers. In digital computers data are represented by a combination of discrete (individual) pulses denoted by 0 and 1. In analogue computers data are represented by a continuously variable physical quantity such as voltage or angular position. Hybrid computers, as the name suggests, combine properties of both digital and analogue computers. The vast majority of computers used in geography are of the digital type and this is reflected in the discussion in this book which concentrates on them. It is possible to represent analogue data, such as river discharge or traffic flows, in digital form by breaking the continuous signal into a series of contiguous discrete data items (see Chapter 9). Digital processing is preferable since for most geographical applications it is usually faster and, at present, it is more sophisticated.

THE STRUCTURE OF COMPUTERS

To understand computers more fully it is necessary to examine their basic structure. This can be achieved by examining how a computer can be used to solve a simple geographical problem, such as finding which of a number of towns is furthest from the city of Plymouth, UK. This problem can be solved manually by measuring the distance on a map to each of the required towns and writing the distances down as a list. To find the longest distance requires working through the list comparing each distance with the previous largest. If the current distance is larger then it becomes the largest. At the end of the list the largest distance is written down as the answer.

A computer could be used to solve this problem in the following way:

1. Input the data (the distances from Plymouth to each town)
2. Store the data in a memory unit (otherwise they will be lost)
3. Compare the distances arithmetically using a logic unit (to work out which town is furthest from Plymouth)
4. Output the answer (otherwise no one will know which is furthest)
5. Use a control unit to oversee these operations (so that they occur in the correct order and at the correct time)

Although five separate operations are required, they are actually carried out by three parts of a computer (Fig. 11.1). One part of a computer normally houses the device for *input* (for example a keyboard), a second the device for *output* (for example a screen) and a third, called the *central processor unit*, houses the arithmetic logic unit, the memory unit and the control unit. These physical parts of a computer are referred to as *hardware*.

In addition to hardware, to solve problems computers require *data* and instructions about how to process the data to find the answer. The total set of instructions used to perform a given task is called a *program* and is referred to as *software* (see Chapter 12).

Central Processor Unit

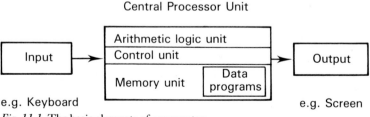

Fig. 11.1. The basic elements of a computer.

THE DEVELOPMENT OF COMPUTERS

Today computers are developing extremely rapidly and any definition of their basic structure and applications must make allowance for this fact. The concept of the computer is generally accredited to Charles P. Babbage, who in 1833 had the idea of developing a machine which he called an Analytical Engine. Babbage set out to build a single machine, using the mechanical technology of the day, which could solve a variety of problems whenever called upon to do so. Unfortunately, only one part of the Analytical Engine was completed, principally because Babbage was restricted to the mechanical technology of the day.

In the following decades the computer evolved relatively slowly. By 1890 Hollerith had developed the Amazing Tabulating Machine to process the data from the American census of that year. Although essentially mechanical, Hollerith's machine made use of electricity to process punched paper cards on which data were coded. In 1925 Bush of the Massachusetts Institute of Technology (MIT) built a large-scale calculating machine, which was also partially electrical.

It was during the 1930s and 1940s that computer development began to gather pace. Zuse developed the first program-controlled computer, the Z1, in 1938. Aiken developed the first electromagnetical computer, the Harvard Mark I, in 1944. This was 55 ft long, 8 ft high and contained almost a million individual components. These machines paved the way for the first general-purpose electronic computer, ENIAC (Electronic Numerical Integrator And Calculator), in 1945. ENIAC contained over 19,000 valves, used vast amounts of electricity, was very large and (by today's standards) had a very small amount of memory. Nevertheless, it was programmable and comparatively fast. In 1951 the first commercial computer, the Ferranti Mark I, was sold in Britain and the Lyons Electronic Office (LEO), the first office computer, was built. Computers were now no longer mechanical work horses ploughing through set tasks, they were the forerunners of today's systems, which are capable of performing quickly a wide variety of operations.

Computer scientists often describe the development of computers as a series of generations, reflecting major technological breakthroughs (Fig. 11.2). First generation computers of the 1940s, such as ENIAC, were based on vacuum tube technology. They tended to be very large and expensive and were short lived and unreliable (nevertheless they were considerably better than anything which had gone before). The second generation of computers were based on transistors. Although transistors were developed in 1947, they were not generally available in computers until 1956. Computers built with transistors were on the whole smaller, more reliable, faster and used less power.

Generation	Electronic Component	Advantages	Disadvantages	Comments
1st generation 1944–55	Vacuum tubes	Vacuum tubes were the only electronic components available.	Large-size. Generated heat. Air-conditioning required. Unreliable. Constant maintenance.	Manual assembly of individual components into a functioning unit.
2nd generation 1956–63	Transistors	Smaller size. Less heat generated. More reliable. Faster.	Air-conditioning required. Maintenance.	As above
3rd generation 1964–70	Integrated circuits	Even smaller size. Even lower heat generation. Less power required. Even more reliable. Faster still.	Initially, problems with manufacture.	Less human labour at assembly stage, therefore cheaper.
4th generation 1971–	Large scale – very large scale integrated circuits	No air-conditioning. Minimal maintenance. High component density. Cheapest.	Very large scale currently less powerful than small scale integrated circuits.	As above

Fig. 11.2. Generations of computer hardware.
Source: Shelly and Hunt 1984 with additions.

The biggest limitation of the first two generations of computers was their cost. They were expensive principally because the components had to be wired together by hand. The breakthrough, which led to a reduction in computer price and evolution of third generation computers, was the ability to combine several computer circuits together on a

(a)

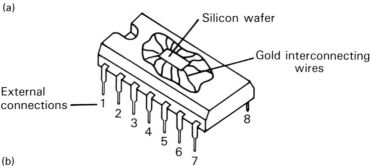

Silicon wafer

Gold interconnecting wires

External connections —— 1

2
3
4
5
6
7
8

(b)

Fig. 11.3. A computer chip. (a) Bottom and top views; (b) Diagram with part of the epoxy resin cut away,

single piece of silicon. These new components were called integrated circuits. They were relatively cheap because they were small and, more importantly, they could be built by machines. Integrated circuits were invented in the late 1950s, but it was 1964 before they were used widely in computers. In the early days of integrated circuits only about ten components could be combined on a surface 5 mm square. This became known as small scale integration (SSI). Subsequent developments have produced medium scale, large scale and now very large scale integration (VLSI), with over 1 million components on an area of 5 mm square by 1 mm thick. Computers using large scale and very large scale integration are customarily called fourth generation computers and have been in use since the early 1970s. Compared to the earlier generations of machines they are cheap, small and reliable.

Today, silicon semiconductor integrated circuits are referred to as *chips*. For convenience chips are mounted in a block of epoxy resin and have metal legs through which to communicate (Fig. 11.3).

MAINFRAME COMPUTER, MINICOMPUTER AND MICROCOMPUTER

Digital computers are often classified into three broad groups on the basis of their relative cost and size, namely, the mainframe computer, the minicomputer and the microcomputer. Fundamentally all three types are similar and, increasingly, the boundaries between them are becoming blurred. First and second generation computers were all *mainframes*. Today mainframes based on more recent technology are common in large business organizations and universities. Typically they are very expensive, require around ten personnel to maintain and run them in special air-conditioned rooms and can support over fifty users at any one time. Very large and powerful mainframe computers fitted with specialist hardware peripherals to increase their processing speed and storage capacity are often called supercomputers.

Minicomputers were developed in the early 1960s, when smaller organizations wanted computing facilities, but did not have sufficient resources for a mainframe. Today most universities, polytechnics and colleges have at least one minicomputer, as do several geography departments. They usually require less than five personnel to run them and can support up to fifty users at the same time.

Microcomputers evolved as a result of large scale integration in the early 1970s. The first microcomputers to have a significant impact on geography were released in 1977. Typically microcomputers are very cheap, support one or at most a few users and require relatively little special maintenance. One of the features of microcomputers is their great variability. Some microcomputers (like that which forms the basis of the soil profile recorder described in Chapter 2) are designed to be portable. Others are specially designed for word processing, graphics and sound. There are also very fast microcomputers which can store large amounts of data and support more than one user simultaneously.

THE ELEMENTS OF A COMPUTER

As a consequence of the great variability in their design and function, it is difficult to produce a succinct, all-embracing definition of what constitutes a computer. However, a basic functional computer unit, suitable for general usage in geography, would normally comprise a processor, keyboard, screen and disk drive. For specific applications in geography, a variety of peripheral devices need to be added to a basic system. These elements will be discussed in more detail below.

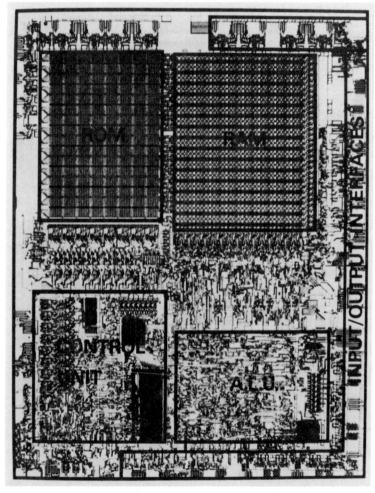

Fig. 11.4. A processor on a single chip. The chip shown is less than 5 mm square by 1 mm thick.
Source: Shelly and Hunt 1984.

Processor

In small cheap microcomputers the three components of a processor (arithmetic unit, memory unit and control unit), together with an input/output interface (to allow communication with other chips), may be on a single chip called a *microprocessor* (Fig. 11.4). Frequently, however, in larger computers they are on more than one chip. The separate chips are mounted on a plastic circuit board and are linked together by metal conductor strips. An individual computer may have the functions of the processor on many chips, on more than one printed circuit board. Figure 11.5 shows a multi-chip, multi-board computer. The most common arrangement is for separate chips to contain the arithmetic logic unit and control unit (this chip is usually called the processor) and the various memory units (see below).

The type of processor, because of its central role, has a great influence on the speed and precision of a computer. In addition, the type of processor is important since some computer programs, especially those written for microcomputers, are written for one or at most a few processor types.

Computer memory is available in a number of different forms each of

Fig. 11.5. The inside of an Apple II multi-board computer.

which performs a slightly different function. *Random Access Memory* (RAM) is the title given to the memory unit in computers which stores users' data and programs. It is sometimes called read/write memory, because it is possible to 'write into' and 'read out from' RAM. When the power is switched off all the information in RAM is usually lost. RAM can be located on a processor chip or, as is more commonly the case, on one or more separate RAM chips. *Read Only Memory* (ROM) is similar in many respects to RAM. The main differences are, first, the contents of ROM are not lost when the power is switched off, and second, it is not possible to write data into ROM and so erase the existing contents. The contents of ROM are usually entered when chips are manufactured. ROM typically contains operating systems, languages and package programs (see Chapter 12), which need not and should not be altered in the normal course of a computer's life. PROM, EPROM and EAROM are all variations of ROM. *PROM (Programmable Read Only Memory)* was introduced to allow specialist programmers to program their own ROMs. A PROM chip is a memory chip into the circuits of which users can load a set of instructions. Once the instructions are loaded, the input wire is cut and the contents of PROM cannot be changed. A second variation of ROM is *EPROM (Erasable Programmable Read Only Memory)*. The difference between EPROM and PROM memory is that the contents of EPROM can be erased several times using a beam of ultra-violet light and then reprogrammed. Although initially more expensive than ROM and PROM, EPROM may work out cheaper in the long run if the programs stored in computer ROM frequently need to be updated. A third variation of ROM is *EAROM (Electrically Alterable Read Only Memory)*. In EAROM individual bits can be reprogrammed using electrical charges. This often reduces the amount of reprogramming when only a small part of the instructions need changing.

The input/output interface of a processor is simply the link between its functional components (arithmetic logic unit, control unit and memory unit) and the other parts of a computer. All data entering and leaving a processor must pass through this interface.

Keyboard

The basic input device for a computer is the keyboard. A computer keyboard is essentially the same as a QWERTY typewriter (the first six keys on the top left hand side are Q-W-E-R-T-Y) with the addition of a few extra keys, such as a RETURN key to send information into the processor for action and RESET to start the computer again from scratch. Many computer keyboards are also fitted with an extra keypad of numbers for rapid input of numerical information.

Screen

The basic output device of a computer is the screen (also called the VDU (Visual Display Unit), display or monitor. Usually the screen is a Cathode Ray Tube (CRT) similar to a television tube (Fig. 11.6), though some special graphics screens operate differently. Pictures are displayed on the screen as a series of illuminated dots. Each dot is an area of phosphor illuminated by a beam of electrons which continuously scans the screen. The computer controls the scanning movement of the electron beam across the screen using guiding plates, turning it on and off to illuminate certain dots. The pattern of movement of the electron beam across the screen is called a raster scan. The scan pattern is usually designed such that each dot is passed about 30 times a second on alternate scans. Computer screens vary in size, usually they are about 12 in (30 cm) wide by about 10 in (25 cm) high, though computers can be connected to 20 in (508 mm) wide and larger screens.

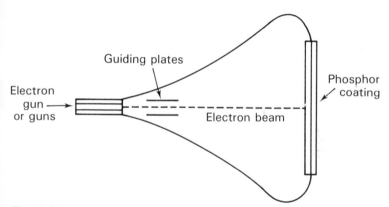

Fig. 11.6. The fundamental components of a computer screen.

Storage

Geographers tend to work with large quantities of data, often more than can be held in the RAM of a computer. The addition of some type of supplementary storage (also called backing store) device is therefore necessary for many applications. There are several types of storage device available of which the most widespread is the disk drive.

Many of today's computers either come with, or have facilities to add on, at least one *floppy disk drive*. Floppy disks are circular and are made of plastic-coated metal oxide enclosed in a protective plastic sleeve (Fig. 11.7). They are a little under 1 mm thick and come in diameters of 8 in

Fig. 11.7. A 5¼-in (134-mm) microcomputer floppy disk.

(203 mm) and 5¼ in (134 mm). More recently 3½ in (74 mm) micro-floppy disks enclosed in rigid plastic cases have been marketed.

The entire disk (including the sleeve) slots into a hole in a floppy disk drive. A small hole in the sleeve allows the read/write head in a disk drive to access the disk to store/retrieve data. Since the read/write head can be positioned quickly over different parts of the disk, storage and retrieval is relatively fast. Typical rates of storage capacity are 100,000 to 500,000 characters per side. Data are stored magnetically on disks in concentric rings, called tracks, which are subdivided into blocks (often referred to as sectors). The block is the basic addressable unit. Disks may have between forty and ninety-six tracks and they can be either single- or double-sided. The most common formats are forty and eighty track (also called single and double density), single-sided or double-sided disks.

The actual mechanism of storage is controlled by a computer Disk Operating System (DOS), a piece of software provided by the computer manufacturer (see Chapter 12). The DOS is, for example, responsible for organizing the blocks of data and for keeping a catalogue of file-names on each disk.

Recent reductions in the price of memory and improvements in computer power have led to increases in the use of Winchester disks in computer systems. A Winchester disk is a type of *hard disk* enclosed in a hermetically sealed chamber. The chamber is filled with lubricant to enable the disk read/write head to take off and land on the disk without causing damage. Winchester disks of $3\frac{1}{2}$ in (74 mm), $5\frac{1}{4}$ in (134 mm) and 8 in (203 mm) diameter, with up to 40 megabyte capacity, are popular and larger versions are also available. In general they are faster, more reliable, and can store more data than floppy disks. However, they are more expensive both to buy and repair.

Peripherals

A peripheral is any device added to a basic computer system. Although input, storage and output devices (such as keyboards, disk drives and screens) are considered essential elements of a computer system, they may also be purchased as peripherals. Many computers are sold as prepackaged units consisting of a keyboard, processor, disk drive and screen; others may comprise a processor only. These options are available so that manufacturers can keep the price of computers as low as possible and so that users can choose the type of devices best suited to their particular applications.

The most common peripheral devices, which can be attached to computer processors, and which are relevant to geography are shown in Fig. 11.8. They are grouped according to their use for input, storage, manipulation and output. Many of these devices are discussed in other chapters, for example, the digitizer, joystick and plotter in Chapter 5 and modem in Chapter 9. A short description of all these devices may be

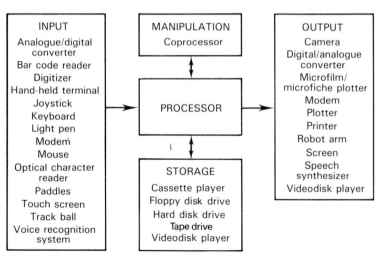

Fig. 11.8. Some computer peripheral devices.

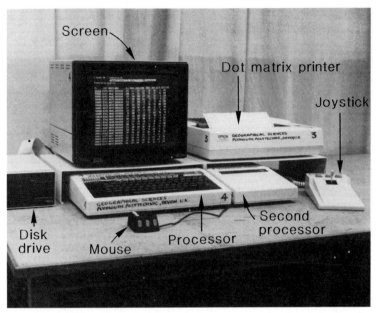

Fig. 11.9. An Acorn microcomputer with a selection of associated peripherals.

found in the Glossary at the back of this book. An Acorn microcomputer and a range of peripherals is shown in Fig. 11.9.

CONCLUSION

This chapter has presented a more detailed description and explanation of the characteristics and operation of the most important computer hardware components. The discussion has been organized around the basic components of computers, namely, the processor and input, output, storage and peripheral devices. The characteristics of computer hardware are important for all computer applications. For example, the type of processor may influence the type of software which can be used in an application, the speed at which it will run and the precision of any answers. The advent of low cost powerful computer hardware and high quality software, which can be easily transferred between machine types, has tended to reduce the significance of hardware characteristics in recent years.

FURTHER READING

Letcher, P. (1985) *Which Peripherals? How to Choose Them, How to Use Them*. Chapman and Hall/Methuen. (A good readable book about most common peripherals.)

Megarry, J. (1985) *Inside Information: Computers, Communications and People*. British Broadcasting Corporation. (An up to date and readable technology book which is well recommended.)

Shelley, J., Hunt, R. (1984) *Computer Studies: a First Course* 2nd edn. Pitman. (A very good computing book which provides a readable introductory account of hardware.)

Walsh, V. (1985) *Computer Literacy: a Beginner's Guide*. Macmillan. (Probably the best introductory book on computers.)

Zorkoczy, P. (1985) *Information Technology: an Introduction* 2nd edn. Pitman. (A well produced introductory text which deals with the basic principles and applications of information technology.)

12

Computer software

This chapter deals with the various aspects of computer software important to geographers. The types of computer software are described and the generations of software languages are discussed, starting with machine code and assembly language, followed by general and specific purpose third generation languages (such as BASIC and FORTRAN) and fourth generation languages. Some example BASIC programs demonstrate the application of programming to solving geographical problems.

Software is a term used to describe all programs which can be used on a computer system. Under this general title come assemblers, compilers, interpreters and operating systems all of which are discussed below. Also included are programs and large multipurpose programs called packages, several examples of which are discussed below and in other chapters. It is because computers can be programmed that they can be used to perform such a wide variety of useful tasks. A computer without software is virtually useless since it cannot be put to work. Such is the importance of software at the present time that many computers are bought and sold, not on the specifications of hardware, but on the ability to run popular software.

People communicate with computers, just as they do with each other, using various *languages*. Computer languages are classified on the basis of whether they are machine-orientated (low level) or problem-orientated (high level). Early computers used machine-orientated languages, but because these were difficult for programmers to understand and use, more problem-orientated languages have been developed.

Level	Generation	Characteristics	Date of Introduction
High ↑	Fourth	Faster, more powerful and specific high level languages eg dBASE and TELL-A-GRAF	late 1970s
	Third	General and specific purpose high level languages eg BASIC and FORTRAN	late 1950s
	Second	Assembly languages	early 1950s
Low	First	Machine code	1940s

Fig. 12.1. Generations of computer programming languages.

These developments occurred in steps and so are often referred to as generations of languages, somewhat analogous to the generations of hardware development described in the previous chapter. At the present time four generations of computer languages are recognized (Fig. 12.1). Computers usually have the facility to program in both high and low level languages, although most users are only aware of the higher levels.

OPERATING SYSTEMS

An operating system is a special piece of software that controls all the individual tasks of a computer, such as editing programs, saving programs and data on disk, and printing programs. Some computers can run more than one operating system (but not at the same time), which makes them very versatile. All computer users have to make use of operating systems and so the range of commands available and their ease of use are important.

MACHINE CODE

Computers really work in only one language which is called machine code. Programs written in other languages must first be translated into machine code before they will run. Such translation is undertaken by special programs, called assemblers, compilers and interpreters (see below).

Computers actually manipulate instructions and data as BInary digiTS (*bits*). In hardware terms a bit is a two-state electronic circuit which can be switched on or off using an electric pulse. In software terms a 1 is used to send an electric pulse to switch a bit on and a 0 to switch it off. This two-state mode of operation uses the logic of the *binary* system. As there are only two characters (0, 1), the binary system is said to work in base two. The decimal system, which we use every day, has ten digits (0123456789) and works in base ten.

In the binary system the right hand digit represents 2 to the power 0 (2^0), the second from the right 2^1 and so on (Fig. 12.2). The decimal equivalent of a binary number is found by adding together the value represented by each bit switched on (that is with a 1 in it). Most computers use one of two standard 8-bit coding systems for storing numbers and other characters. The most widely used is the *ASCII* (American Standard Code for Information Interchange) system. A variation of ASCII called EBCDIC is used by IBM.

A group of 8 bits is referred to as a *byte* (abbreviated to b). The term *kilobyte* (kb) is used for 2^{10} (1,024) bytes, so 32 kb = 32,768 b. The term *megabyte* (mb) is used for 2^{100} (1,048,576) bytes = 1,024 kb, so 4 mb = 4,194,304 b. The number of bits parcelled together to give a standard unit for handling a single character (number, letter etc.) is called a *word*. Many microcomputers have 8- or 16-bit words, but 32 and larger word sizes are more common in minicomputers and mainframe computers. These terms are important since they are measures of the power of computers.

The early computers of the 1940s could only be programmed using machine code. It was, and still is, difficult to learn and tedious to use, since the binary codes are difficult to remember. As a result it is easy to make mistakes and difficult to correct them. Moreover, machine code is very machine-specific. There are, however, two main advantages of

	msd							lsd
Digit number	8	7	6	5	4	3	2	1
	1	0	1	0	1	0	0	1
Power	2^7	2^6	2^5	2^4	2^3	2^2	2^1	2^0
Value if 1	128	64	32	16	8	4	2	1
Decimal equivalent = 169	128		32		8			1

Fig. 12.2. The binary representation of numbers; lsd = least significant digit; msd = most significant digit.

writing programs in machine code. First, they run faster than equivalent programs written in high level languages such as BASIC. This is because machine code programs do not have to be converted into binary each time they are run. Second, specialist operations, which are not available in high level programming languages, can be programmed.

ASSEMBLY LANGUAGES

As a result of the problems experienced in using machine code, a second generation of computer languages has evolved, called *assembly languages*. These allow instructions to be represented in a more meaningful way, for example, ADC may be used to add together data and JSR may be used to jump to a subroutine (see the section on BASIC for an explanation of this term). To make a computer carry out the instructions in an assembly language program, the instructions must first be assembled into machine code by an *assembler*. An assembler is simply a program, written in either a high or low level language, which converts assembly language programs (called source code) into machine code (called object code). The assembler then places the machine code in the appropriate area of memory, together with any other specified data or instructions. A further command can then be used to make the processor carry out the machine code instructions.

THIRD-GENERATION LANGUAGES

Most people's concept of programming encompasses the idea of programming in one of the high level languages. These languages are orientated towards problem solving. They use decimal numbers, characters of the English alphabet and a vocabulary of reserved words, such as PRINT, LIST and FORMAT. In addition, operators, such as + − * / AND OR = ><, are used to perform arithmetical, relational and logical operations. Thus high level languages are much easier to use than low level languages. They are designed to organize human ideas and logic into such a form that computers can process them. The first high level language, FORTRAN, came into use on mainframe computers in the middle 1950s. FORTRAN, and the other high level languages developed from the late 1950s onward, are referred to as third generation languages and represent a further abstraction from the binary logic of computer machine code.

There is insufficient space here to recount anything other than the basic elements of programming using third generation languages. In any case many good books are available on the subject. The book by Unwin and Dawson (1985) is particularly relevant to geographers and environmental scientists.

General-purpose third-generation languages

General purpose third generation high level languages can be used to solve a wide variety of problems and have a flexible and versatile structure. Space does not permit lengthy discussion of all the languages, since there are well over fifty. The intention is to discuss BASIC, as an example of a third generation general purpose language, with emphasis on how it may be used to solve geographical problems. This is followed by brief details of the other two main languages used by geographers: FORTRAN and Pascal. Further details and a summary of several other languages can be found in Marshall (1983), Martin (1984) and Moller (1984).

BASIC

BASIC is by far and away the most widely used programming language on computers, indeed it comes as standard with virtually all systems. It was devised by J. G. Kemeny and T. E. Kurtz of Dartmouth College, USA and released in 1965. BASIC has many similarities with FORTRAN from which it was developed. It was designed with an integral editor for use on an interactive time shared system. In such a system a computer supports several users at the same time, by switching from one to another in a fixed sequence, processing a certain number of instructions on each occasion. BASIC was also intended to be easy to learn and teach, hence the name BASIC which stands for Beginners All purpose Symbolic Instruction Code. There are many versions of BASIC (known as dialects or, more euphemistically, as 'Street BASICS') and whilst no official standard exists, many use a similar kernel of commands (Kemeny and Kurtz 1985).

Any program written in BASIC consists of a series of statements' one on each line of a program. The statements are numbered sequentially, to keep them in order, and to allow movement around a program. The simple BASIC program called POPGB (Fig. 12.3a), performs the four fundamental tasks of a computer, namely input, storage, manipulation and output. The command INPUT is used to input data from the keyboard into memory. The populations of England, Scotland, Wales and Great Britain (once it is calculated) are stored in memory locations, identified in this program as E, S, W and GB respectively. The data are manipulated on line 100, where the populations of England, Scotland and Wales are added together and stored in a memory location identified as GB. The command PRINT is used to output information to the

(a)
```
 10 REM----A SIMPLE BASIC PROGRAM POPGB----
 20 REM---INPUT---
 30 PRINT"WHAT IS THE POPULATION OF ENGLAND IN MILLIONS"
 40 INPUT E
 50 PRINT"WHAT IS THE POPULATION OF SCOTLAND IN MILLIONS"
 60 INPUT S
 70 PRINT"WHAT IS THE POPULATION OF WALES IN MILLIONS"
 80 INPUT W
 90 REM----CALCULATION----
100 GB = E + S + W
110 REM----OUTPUT---
120 PRINT"THE POPULATION OF GREAT BRITAIN= "
130 PRINT GB
140 END
```

(b)
```
WHAT IS THE POPULATION OF ENGLAND IN MILLIONS
?47
WHAT IS THE POPULATION OF SCOTLAND IN MILLIONS
?5
WHAT IS THE POPULATION OF WALES IN MILLIONS
?3
THE POPULATION OF GREAT BRITAIN=
        55
```

Fig. 12.3. A simple BASIC program, POPGB, which calculates the population of Great Britain. (a) Program listing; (b) transcript of a run.

screen. Two types of PRINT output are used (Fig. 12.3b). On lines 30, 50, 70 and 120 the PRINT command is used to output text to the screen. The text is used to aid the readability of the second type of output, that is the output of the contents of various locations of memory. For example, the command on line 130 means PRINT on the screen the contents of the memory location identified as GB.

The program also contains REM and END commands. The REM command is used to break the program into sections and to insert REMarks to improve readability. When a computer encounters a REM command it ignores everything else which follows on that line. The END command, as might be expected, marks the end of the program.

The statements which make up a program are simply typed into a computer, with a RETURN (press of the return key) used to end a line. A computer will not carry out the instructions until the RUN command is issued. When a computer RUNs a BASIC program it reads each line, converts it to machine code and processes it. This operation is undertaken by an *interpreter*, which is itself a special program in the memory of a computer. Since each line must be converted into machine code every time a program is run, interpreted languages tend to execute comparatively slowly (in comparison to assembled and compiled languages). However, editing small programs written in interpreted languages can be simpler and quicker. For those who require programs to run quickly, compiled versions of BASIC are available.

Data used by a BASIC program are stored in various memory locations, identified in a program by a label, or variable name to use the BASIC term. In BASIC it is important to differentiate between numerical variables (such as 11.12, 2,701 and 000.3) and string or alphanumerical variables (such as ROBSN, BOYCT and FNW862Y). To manipulate both numerical and string data, BASIC has arithmetical and relational operators and a number of mathematical functions. These allow complex mathematical formulae to be incorporated into programs such as:

$$T = R * SQR ((N-2)/(1-(R^2)))$$

This should be recognizable as a Student's *t* test on a correlation coefficient which has previously been calculated and stored as variable R, where N is the number of observations in the data set. The formula for calculating Student's *t* is normally written as:

$$t = r * \sqrt{\frac{n-2}{1-r^2}}$$

In BASIC it is possible to repeat operations a number of times by using FOR-NEXT loops. The programming of a FOR-NEXT loop is more complicated than the operations described so far, but understanding it in principle is fairly straightforward. A FOR-NEXT loop simply allows a section of a BASIC program to be repeated a given number of times. The FOR command indicates the beginning and the NEXT command the end of a loop. An example of a FOR-NEXT loop to INPUT five numbers into a program and add them up is shown in Fig. 12.4. The FOR-NEXT loop has a counter (I) assigned to it on line 20. Each time line 20 is executed, I is incremented by 1. When the operations in the loop have been carried out five times the program stops at line 60. In this program data are input at line 40 and stored in variable *X*. The sum of all the *X* values is stored in variable *T*.

BASIC also has facilities to control movement around programs by using GOTO, IF-THEN and GOSUB commands. These commands are useful where, for example, the operations of the second part of a

```
20 FOR I=1 TO 5
30 PRINT"INPUT NUMBER"; I
45 INPUT X
50 T=T+X
60 NEXT I
```

Fig. 12.4. A simple example of the use of a FOR-NEXT loop in BASIC. Note that this is an incomplete program.

```
10 REM ----BASIC PROGRAM----
20 REM  TO CALCULATE MEAN
30 N=0
40 T=0
50 M=0
60 PRINT"ENTER OBSERVATION"
70 INPUT X
80 IF X=-99 THEN 120
90 N=N+1
100 T=T+X
110 GOTO 60
120 M=T/N
130 PRINT"N= ";N
140 PRINT"MEAN= ";M
150 END
```

Fig. 12.5. A simple example of the use of IF-THEN and GOTO commands in BASIC.

program are dependent upon some calculations in the first part. IF-THEN and GOTO commands can be used to control a simple program which calculates the mean of several observations (Fig. 12.5). The program prompts for input on line 60 and stores the data in variable X. Line 90 keeps a count of the number of observations (N) and line 100 calculates the sum total (T) of the observations. The GOTO command on line 110 ensures that this process is repeated unless a value of -99 is entered. If a value of -99 is entered, the IF-THEN command on line 80 terminates input by making the computer execute lines 120 – 150, which calculate the mean (M), print the results and end the program.

The command GOSUB is used in situations where it is necessary to perform the same operation, or group of operations, on more than one occasion in a program. For example, it may be necessary to calculate the standard deviation of several sets of data within a single program. Rather than code up the operation each time it is required in a program, it is simpler and shorter to code up the operation once as a subroutine. Each time the operation is required the subroutine can be called and the statements in the subroutine executed. When the operation in the subroutine is complete the computer returns to the main program and executes the statement on the line below the line with the GOSUB command.

The final aspect of the BASIC language to be discussed here is the *array*. The idea of variables as individual storage locations has already been introduced. Sometimes it is desirable to store and manipulate several related data items together, as in the case of the soil data shown in Fig. 12.6a. It is tidier and, because of the way BASIC is designed, easier to manipulate the data as an array. An array can be thought of as a matrix where the number of rows is equal to the number of observations and the number of columns is equal to the number of variables. An array

(a)

Observation	Soil Type			
	1 Brown Earth	2 Podzol	3 Gley	4 Peat
1	3.6	8.4	25.2	89.1
2	2.9	9.5	21.1	85.6
3	2.8	8.3	19.7	79.1
4	4.7	10.2	19.7	80.0
5	4.4	9.8	22.3	90.3

(b)

$$S \begin{pmatrix} 1,1 & 1,2 & 1,3 & 1,4 \\ 2,1 & 2,2 & 2,3 & 2,4 \\ 3,1 & 3,2 & 3,3 & 3,4 \\ 4,1 & 4,2 & 4,3 & 4,4 \\ 5,1 & 5,2 & 5,3 & 5,4 \end{pmatrix}$$

Fig. 12.6. A simple example of an array in BASIC. (a) The organic matter
content of five samples from four soil types; (b) an array called S,
which could be used to store the data, showing the subscripts for
each data item.

is given a single name and each element is referred to by subscripts. The
array containing the soil data could be called S (Fig. 12.6b). It has 5 rows
and 4 columns (two dimensions) and is, therefore, referred to as S(5,4).

The DESTATS program (Fig. 12.7) utilizes many of the concepts
and commands introduced above. Lines 100 – 160 contain instructions

(a)
```
 10 REM ----A BASIC PROGRAM----
 20 REM        DESTATS
 30 REM
 40 REM     DAVID J. MAGUIRE
 50 REM
 60 REM ----DEFINITION----
 70 DIM X(1000)
 80 T=0
 90 SDT=0
100 REM ----INPUT----
110 PRINT"HOW MANY OBSERVATIONS ARE THERE?"
120 INPUT NUMBER
130 FOR I=1 TO NUMBER
140 PRINT"ENTER OBSERVATION ";I
150 INPUT X(I)
160 NEXT I
170 REM ----CALCULATION----
180 REM MEAN
190 FOR I=1 TO NUMBER
200 T=T + X(I)
210 NEXT I
```

```
220 MEAN=T/NUMBER
230 REM MIN & MAX
240 MIN=99999:MAX=-99999
250 FOR I=1 TO NUMBER
260 IF X(I) < MIN THEN MIN=X(I)
270 IF X(I) > MAX THEN MAX=X(I)
280 NEXT I
290 REM STANDARD DEVIATION
300 FOR I=1 TO NUMBER
310 SDSUM=((X(I)-MEAN)^2)
320 SDT=SDT+SDSUM
330 NEXT I
340 SD=SQR(SDT/NUMBER)
350 REM ---OUTPUT---
360 GOSUB 1000
370 PRINT"TOTAL= ";T
380 PRINT"MEAN= ";MEAN
390 PRINT"MINIMUM= ";MIN
400 PRINT"MAXIMUM= ";MAX
410 PRINT"STANDARD DEVIATION= ";SD
420 GOSUB 1000
430 END
1000 REM ---- SUB1 - PRINT ----
1020 PRINT
1030 PRINT"=========================="
1040 PRINT
1050 RETURN
```

(b)
```
HOW MANY OBSERVATIONS ARE THERE?
?6
ENTER OBSERVATION 1
?89.1
ENTER OBSERVATION 2
?85.6
ENTER OBSERVATION 3
?79.1
ENTER OBSERVATION 4
?80.0
ENTER OBSERVATION 5
?90.3
ENTER OBSERVATION 6
?100.9


==================================

TOTAL= 525
MEAN= 87.5
MINIMUM= 79.1
MAXIMUM= 100.9
STANDARD DEVIATION= 7.30730229

==================================
```

Fig. 12.7. A BASIC program called DESTATS which demonstrates several of the features of the BASIC Language. (a) Program listing; (b) transcript of a program run using hypothetical data.

to input up to 1,000 observations and store them in the array X(I). The mean, minimum, maximum and standard deviation of the data are calculated on lines 170 – 340. A short subroutine (lines 1000 – 1050) used to enhance the readability of the output, is called on line 360.

FORTRAN

FORTRAN was developed in the mid 1950s and was the first high level programming language. The name FORTRAN is a contraction of FORmula TRANslation, which is a pointer to the main uses of the language for solving numerical problems. It was taught to many geography undergraduates in the 1970s and is still in frequent use in many departments. Even today it remains the most widespread language for scientific and engineering programming, for several reasons. Many people have been using FORTRAN for a number of years and, because they are familiar with its structure and logic, are reluctant to change. A large amount of software is available, which can be incorporated into FORTRAN programs, including the NAG (Numerical Algorithm Group) and GINO (Graphical INput Output) subroutine libraries. There is an international standard for FORTRAN, specified by the American National Standards Institute (ANSI), that facilitates language portability between machines.

FORTRAN has many similarities to BASIC, the major difference being that FORTRAN is a compiled language. This means that before a FORTRAN program can be run the whole program must be compiled into machine code, using a special computer program called a *compiler.* The machine code (called the object code) can then be kept and used many times. However, because machine code is difficult to read and understand, changes are normally implemented on the original high level code (called the source code) which is then recompiled. FORTRAN is, generally speaking, more difficult to learn and use than BASIC, as a result of its requirement for precise syntax and ordering. Principally for this reason, BASIC is usually preferred for undergraduate teaching.

Pascal

Pascal was devised in 1970 by N. Wirth of Zurich, and named after the French mathematician Blaise Pascal. It was created with the objective of teaching programming and soon became popular. In 1973 it was adopted by the University of California San Diego (UCSD), who have since implemented it on a wide range of computers. Like BASIC several different dialects of Pascal are available, but the UCSD implementation seems to have been adopted as the *de facto* standard. The book by James

(1983) provides a guide to several of the dialects. Pascal is taught to a number of geography students, especially in the USA.

Pascal programs have a formal structure, with statements organized into two sections: a definition section and an instruction section. Further structure is gained by enclosing blocks of code within BEGIN and END commands. There are no line numbers in Pascal, indeed there is no concept of a line. Each statement is simply separated from the next by a semicolon.

Specific-purpose third-generation languages

Specific purpose high level languages are even more problem-orientated than general purpose high level languages. Each language possesses distinctive features which reflect the purposes for which it was developed. There are well over a dozen specific purpose third generation high level languages, too many to discuss here. Details of several languages, including ALGOL, BCPL, C, PILOT, POPLOG and PROLOG are described in Marshall (1983) and Moller (1984). As far as geographers are concerned, one of the most useful specific purpose third generation languages is LOGO.

LOGO

In recent years a great deal of attention has been devoted to LOGO. It was developed in the USA in the 1960s under the inspiration of Seymour Papert. LOGO is potentially a very powerful language that is easy to use. At its simplest level LOGO can be used to explore computer graphics by moving a cursor (which is made to look like a turtle) around a screen. It also has facilities for processing lists of data. In geography LOGO can probably best be used to introduce the ideas of programming graphics and devising logical solutions to problems, although more sophisticated applications can be undertaken.

FOURTH-GENERATION LANGUAGES

Third generation languages represented a major step forward in computing. Their natural language key words and mathematical notation made them much easier to understand and use. However, third generation languages require vast amounts of program code for typical large applications and they are designed for use by qualified data processing professionals and experienced academics. Fourth generation languages were created in response to these problems. According to Martin (1984) the objectives of fourth generation languages are:

1. to speed up application building and editing;
2. to generate code with few bugs and minimum debugging problems;
3. to make languages easier to understand and use.

Fourth generation languages are a fast, efficient and easy to learn way of writing program code. They make use of data dictionaries and use commands closely related to natural language. A data dictionary is a store of information about various applications such as screen displays, calculations and output formats, which may be utilized in programs. Many fourth generation languages have facilities for screen-based data entry, querying data bases and generating reports. They are more problem-orientated than the earlier generations of languages and so, increasingly, it is necessary to select a specific language for a specific application.

Fourth generation languages can be classified according to their uses. There are graphics languages such as TELL-A-GRAF and SAS/GRAPH (see Chapter 5); query, report and application generators such as dBASE III (see Chapter 3); decision support languages, that include the spreadsheet language Lotus 1–2–3 (see Chapter 4); and authoring languages such as Microtext.

CONCLUSION

This chapter on computer software, like the previous chapter on hardware, is designed for technically minded geographers who wish to know more about how computers actually operate. The main types of computer software have been discussed, starting with the low level machine-orientated languages machine code and assembly language, followed by the high level problem-orientated languages such as BASIC, FORTRAN and Pascal. Only a few years ago the lack of good quality software meant that all computer users had to be proficient in at least one high level language. However, the availability of good quality, error-free standard applications software, such as that described in earlier chapters, means that the need for geographers to be able to program has been reduced.

FURTHER READING

Alcock, D. (1977) *Illustrating BASIC*. Cambridge University Press. (Good and simple, but gimmicky.)
James, M. (1983) *Pascal for Micros*. Butterworth. (A clear and concise introduction to Pascal.)

Marshall, G. J. (1983) *Programming Languages for Micros*. Butterworth. (A useful overview of high level languages).

Martin, J. (1984) *Fourth Generation Languages. Volume II: Survey of Representative 4GLs*. Savant. (Brief details about nearly 40 4GLs. The introduction about generations of languages is particularly interesting.)

Mather, P. M. (1976) *Computers in Geography: a Practical Approach*. Blackwell. (FORTRAN programming for geographers. Some good examples, but now rather dated.)

Unwin, D. J., Dawson, J. A. (1985) *Computer Programming for Geographers*. Longman. (A good book by geographers, for geographers, on programming using BASIC and FORTRAN.)

13

Conclusions and the future

The aim of this brief final chapter is to bring together and summarize the major themes discussed earlier and to offer some suggestions as to likely future developments in geographical computing.

Computers are essentially an enabling technology: they are tools which enable geographers to improve their efficacy and efficiency in many ways. There are two key aspects of computers which assist geographers in this regard. First, computers can be used to collect and store large quantities of data in an organized manner. Second, such data can be quickly manipulated and presented in a whole host of different ways. It is now possible to use computers to collect data about people and the environment using automatic and semiautomatic data loggers. Data base management systems can be used to organize the storage of these data which can then be made available to many users via local and wide area networks and telephone links. Data extracted from data bases can be manipulated in many ways using statistical analysis, computer cartography, image analysis and geographical information system software. The results of any analyses can be attractively presented with the assistance of word processing software.

Some of the material in this book has been concerned with how computer devices and programs operate and the methods that computers provide for geographers to use. Greater emphasis, however, has been given to the use of computers for investigating substantive geographical problems. Some would argue that it is only when computers are applied to real world problems are they of any real use to the geographer. Computers can be used to investigate a very wide range of substantive geographical problems and it seems worthwhile summarizing a few examples to illustrate this point. Computers have been used to assist with the collection of soil profile data in order to describe the character-istics and distribution of soil types in Britain. They have been used to organize, analyse and map population census data in order to investigate the socio-economic characteristics of the city of Plymouth, UK. They have been used to monitor thermal pollution levels in the Savannah River, South Carolina, USA using satellite remote sensing data. Finally, they have been used to explain and predict variations in the rainfall pattern in Northumberland, UK.

Computers have made an important contribution to teaching as well as research. Computer Assisted Learning (CAL) usually refers to teaching with rather than about computers, though given society's current reliance on computers, it is surely important that geographers should be acquainted with the basic principles of their operation as well as their applications. There are many examples of how computers are being used for CAL. These include developing problem solving skills using role playing simulations, explaining difficult concepts and rein-forcing ideas presented in lectures and practicals with computerized tutorial programs, and encouraging students to explore relationships between phenomena using data analysis programs. There is a good selection of CAL software that can be used by geography lecturers and teachers, but it remains up to individuals to incorporate them into their own courses as they see fit.

During the next few years a number of major technological and methodological developments may be anticipated that will have a signi-ficant affect on geographical computing. The most important advances in hardware include new high capacity data storage devices, based on videodisk and compact disk technology, and very powerful new micro-processors which manipulate instructions very rapidly. The most signifi-cant developments in software are likely to be more widespread adop-tion of fourth generation programming languages and improvements in the way we use computers. Already a number of computers support graphical user interfaces called WIMPS (Window Icon Mouse Pointer Systems).

Looking further to the future, a number of countries have recognized the economic and strategic significance of information technology and have embarked on major research and development projects aimed at improving current computer systems. The first of the large scale nation-al schemes was the Japanese 'fifth generation' project which started in 1982 and was planned to take ten years. In response to this, several other countries embarked on similar schemes of which the most widely publi-cized are probably the American Strategic Defence Initiative (SDI) or 'star wars' project and the British Alvey project. The exact details of all these schemes differ, but a number of themes are common to them all. They all aim to develop new computers which will be faster, more powerful, more reliable, easier to use and cheaper than existing systems. This is to be achieved through initiatives in microelectronics, software development and the human/computer interface.

The initiatives in microelectronics are concerned with further miniaturization of computer chips, leading to ultra large scale integrated circuits, and with radical departures from existing processor designs, allowing computers to process instructions in parallel rather than in series as they do at present. Already these developments are greatly increasing the speed and power of computers and, as a result, will further enhance the scope of geographical computing.

The initiatives in software are concerned with the development of expert systems (also called intelligent knowledge-based systems) using techniques developed in the field of artificial intelligence (Smith 1984). *Expert systems* are computer programs which capture the knowledge of highly skilled experts, usually in the form of a set of rules or lexicon of principles, and make it available to novice users. Even now there are several geographical expert systems in existence (Davis, Hoare and Nanninga 1986; Robinson, Frank and Matthew 1986; Nichol, Briggs and Dean 1987). They are used for deducing the possible presence of minerals, for improving fire management strategies and for making GISs easier to use by adapting themselves to user enquiries.

The initiatives in the human/computer interface are concerned with the input and output of information by speech, graphics, images and documents. Already some primitive devices are available for computer vision and sound recognition. Clearly further developments in this area will make computer systems much easier to use and more acceptable to ordinary people.

I suppose the next stage in the development of computers, the sixth generation, will see machines which not only think for themselves but also argue with themselves and laugh at their own jokes!

Bibliography

Adams, A., Jones, E. (1983) *Teaching Humanities in the Electronic Age.* Open University Press

Alcock, D. (1977) *Illustrating BASIC.* Cambridge University Press

Anderson, M. G. (1982) Modelling hillslope soil water status during drainage. *Transactions of the Institute of British Geographers* 7: 337–53

Armstrong, A. C., Whalley, W. B. (1985) An introduction to data logging. *British Geomorphological Research Group Technical Bulletin* **34**

Bagshaw, E. (1985) Image makers. *Business Computing: the Survival Game. Personal Computer World Special* 68–72

Balchin, W. G. V. (1985) Graphicacy achieves examination status. *Area* 17: 256

Balchin, W. G. V., Coleman, A. M. (1965) Graphicacy should be the fourth ace in the pack. *Times Higher Education Supplement* 5 November

Bale, J. (1981) *The Location of Manufacturing Industry* 2nd edn. Oliver and Boyd

Ball, A. P., Browning, K. A., Collier, C. G. *et al.* (1979) Thunderstorms developing over Northwest Europe as seen by Meteosat and replayed in realtime on a fast-replay colour display. *Weather* 34: 141–7

Barlow, N. D., Dixon, F. G. (1980) *Simulation of Lime Aphid Population Dynamics.* PUDOC (Centre for Agricultural Publishing and Documentation), Wageningen

Barrett, E. C., Martin, D. W. (1981) *The Use of Satellite Data in Rainfall Monitoring.* Academic Press

Batty, M. (1987) *Microcomputer Graphics: Art Design and Creative Modelling.* Chapman and Hall

Batty, M., Longley, P. A. (1986) The fractal simulation of urban structure. *Environment and Planning A* **18**: 1143–79

Bennett, R. J., Wrigley, N. (1981) Introduction. In **Wrigley, N., Bennett, R. J.** (eds) *Quantitative Geography.* Methuen, pp. 3–11

Bickmore, D. (1987) Scientific roles for the new cartography. *The Cartographic Journal* **24**: 56–8

Bishop, O. N. (1983) *Statistics for Biology: a Microcomputer Edition.* Longman

Blakemore, M. (1985) High or low resolution? Conflicts of accuracy, cost, quality and application in computer mapping. *Computers & Geosciences* **11**: 345–8

Boggs Mathews, C., Mathews, M. S. (1985) *Word Processing for the IBM PC & PCjr and Compatible Computers.* McGraw-Hill, New York

Bonczek, R. H., Hosapple, C. W., Whiston, A. B. (1984) *Micro Database Management: Practical Techniques for Application Development.* Academic Press, Orlando, USA

Bond, G. (1984) A data base catalog. *Byte* October: 227–38

Bowlby, S., Silk, J. (1982) Analysis of qualitative data using GLIM: two examples based on shopping survey data. *Professional Geographer* **34**: 80–90

Bracken, I. (1985) Computer-aided cartography with microcomputers: a practical guide to MicroPLOT. *Papers in Planning Research UWIST Department of Town Planning Research* **90**

Bracken, I. (1986) Digitizing principles and applications in micro-computer assisted cartography. *Papers in Planning Research UWIST Department of Town Planning Research* **96**

Bracken, I., Spooner, R. (1985) Interactive digitizing for computer-assisted cartography. *Area* **17**: 205–12

Brodlie, K. W. (1985) GKS – The standard for computer graphics. *Computers & Geosciences* **11**: 339–44

Brown, D. E., Winer, A. M. (1986) Estimating urban vegetation cover in Los Angeles. *Photogrammetric Engineering and Remote Sensing* **52**: 117–23

Burcham, R., Ferguson, I. (1985) *Factory location (FACLOC).* Longman Micro Software

Burges, S., Piercy, N. (1985) Word processing: a question of a hit or a myth. *Times Higher Education Supplement* 20 September

Burgess, M. D., Hanson, C. L. (1983) Microprocessor-controlled precipitation data collection system. *Journal of Hydrology* **66**: 369–74

Burkhardt, H., Fraser, R., Clowes, M., Eggleston, J., Wells, C. (1982) *Design and Development of Programs as Teaching Material. MEP Information Guide 3.* Council for Education Technology

Burrough, P. A. (1986) *Principles of Geographical Information Systems for Land Resources Assessment. Monographs on Soil and Resources Survey* **12**. Clarendon Press

Burt, T., Butcher, D. (1986) Simulation from simulation? A teaching model of hillslope hydrology for use on microcomputers. *Journal of Geography in Higher Education* **10**: 23–39

Byers, R. A. (1983) *Everyman's Database Primer: Featuring dBASE II.* Blackwell

Cable, D., Rowe, B. (1987) *Software for Statistical and Survey Analysis.* Study Group on Computers in Survey Analysis

Carpenter, J., Deloria, D., Morgenstein, D. (1984) Statistical analysis for microcomputers. *Byte* April: 234–64

Carruthers, A. W. (1985) *Introductory User Guide to GIMMS.* GIMMS Ltd

Carter, J. R. (1980) Thematic mapping on low-resolution color CRT terminals. *Harvard Library of Computer Graphics* **9**: 35–40d

Carter, J. R. (1984) Computer mapping: progress in the '80s. *Resource Publications in Geography*. AAG, Washington

Catlow, D. R. (1986) The multi-disciplinary applications of DEMS. In **Blakemore, M.** (ed) *Proceedings of Auto Carto, London Vol 1* Royal Institute of Chartered Surveyors, London pp. 447–54

Census Research Unit (1980) *People in Britain – a Census Atlas.* HMSO

Chalmers, L., Thompson, D., Keown, P. (1979) The use of a computer based simulation model in the geography classroom. *New Zealand Journal of Geography* **67**: 6–9

Chambers, J. M. (1980) Computers in statistics. *American Statistician* **34**: 238–43

Chan, J. M. T., Korostoff, M. (1984) *Teachers' Guide to Designing Classroom Software.* Sage, Beverley Hills

Chorley, R. J., Haggett, P. (1967) *Models in Geography.* Methuen

Chrisman, N. R. (1983) The role of quality information in the long-term functioning of a geographic information system. In **Weller, B. S.** (ed) Automated cartography: international achievements and challenges. *Proceedings of the 6th International Symposium on Automated Cartography*, pp. 303–12

Clarke, S. R., Fisher, P. F., Ragg, J. M. (1986) Soil Profile Recorder: a program to enable the recording of soil profile descriptions in the field. *Computers & Geosciences* **12**: 779–806

Cohen, L., Holliday, M. (1982) *Statistics for Social Sciences.* Harper and Row

Cooke, D. A., Craven, A. H., Clarke, G. M. (1982) *Basic Statistical Computing.* Edward Arnold

Cooke, D. A., Craven, A. H., Clarke, G. M. (1985) *Statistical Computing in Pascal.* Edward Arnold

Cook, K. L. (1984) SLICK – a review. *Teaching Geography* **10**: 39–40

Coppock, J. T. (1964) *Agricultural Atlas of England and Wales.* Faber

Curran, P. J. (1985) *Principles of Remote Sensing.* Longman

Daker, L. (1987) Information handling – QUEST. In **Kent A.** (ed) *Computers in the Classroom.* Geographical Association, pp. 10–15

Dangermond, J. (1983) A classification of software components commonly used in geographic information systems. In **Peuquet, D., O'Callaghan, J.** (eds) *Design and Implementation of Computer-based Geographic Information Systems.* IGU Commission on Geographical Data Sensing and Processing. Amherst, NY

Date, C. J. (1981) *An Introduction to Data Base Systems* 3rd edn. Addison-Wesley

Dawson, J. A., Unwin, D. J. (1976) *Computing for Geographers.* David and Charles

Dawson, J. A., Unwin, D. J. (1984) The integration of microcomputers into British Geography. *Area* **16**, 323–9

Day, C. A. (1984) *Text Processing. Cambridge Computer Science Texts 20.* Cambridge University Press

Degani, A., Lewis, L. A., Downing, B. B. (1979) Interactive computer simulation of the spatial process of soil erosion. *Professional Geographer* **31**: 184–90

Denley, P. (1986) *Word Processing and Publishing: some Guidelines for Authors.* Office for Humanities Communication University of Leicester

Denley, P., Hopkin, D. (1987) (eds) *History and Computing.* Manchester University Press

Dickinson (1973) *Statistical Mapping and the Presentation of Statistics* 2nd edn. Edward Arnold

DoE (1987) *Handling Geographic Information.* HMSO

Dougenik, J. A., Sheehan, D. E. (1975) *SYMAP User's Reference Manual.* Laboratory for Computer Graphics and Spatial Analysis. Harvard, Massachusetts

Dozier, J., Outcalt, S. I. (1979) An approach towards energy balance simulation over rugged terrain. *Geographical Analysis* **11**: 65–85

Drury, S. A. (1987) *Image Interpretation in Geology.* Allen and Unwin

Ebdon, D. (1985) *Statistics in Geography* 2nd edn. Basil Blackwell

Ellison, D., Tunnicliffe Wilson, J. C. (1984) *How to Write Simulations using Microcomputers.* McGraw-Hill

Ettlin, W. A. (1982) *WordStar Made Easy* 2nd edn. Osborne/McGraw-Hill, Berkeley

Fenn, C. R. (1987) Electrical conductivity. In **Gurnell, A. M., Clark, M. J.** (eds) *Glacio-Fluvial Sediment Transfer: an Alpine Perspective.* Wiley, pp. 377–414

Fisher, P. F., Pearson, M. D., Clarke, S. R., Ragg, J. M. (1987)

Computer programs to assist the automation of soil description. *Soil Use and Management* 3: 26–31

Fleming, G. (1975) *Computer Simulation Techniques in Hydrology.* Elsevier, New York

Forer, P. (1984) *Applied Apple Graphics.* Prentice-Hall, Englewood Cliffs, New Jersey

Forrester, J. W. (1969) *Urban Dynamics.* MIT Press, Cambridge, Massachusetts

Fox, P. S. (1984) *List of Geography Microcomputer Software.* Geographical Association

Gardiner, V., Unwin, D. J. (1986) Computers and the field class. *Journal of Geography in Higher Education* 10: 169–79

Garrison, W. L. (1959) Spatial structure of the economy. *Annals of the Association of American Geographers* 49: 232–9

Gee, K. C. E. (1982) *Local Area Networks.* National Computer Centre Publications

Gilchrist, R. (1985) Statistical packages for the IBM PC. In **Barnetson, P.** (ed) *The Research and Academic Users' Guide to the IBM Personal Computer.* IBM UK Ltd, pp. 111–29

Goddard, J., Armstrong, P. (1986) The 1986 Domesday project. *Transactions of the Institute of British Geographers* 11: 290–5

Gray, R. J., Maizel, M. S. (1985) *A Survey of Geographic Information Systems for Natural Resources Decision Making.* The American Farmland Trust, Washington

Green, D., Baker, A. M., Deeth, A. L., Nuzzo, A., Faludi, R. (1985) SAS/GRAPH for cartography: map projections and labelled choropleth maps. *Cartographica* 22: 63–78

Green, N. P. A., Rhind, D. W. (1986) Teach yourself geographic information systems: the design, creation and use of demonstrators and tutors. In **Blakemore, M.** (ed) *Proceedings of Auto Carto, London.* Royal Institute of Chartered Surveyors, pp. 327–35

Grelot, J.-P. (1986) Archaic data models or hardware as a concept killer. In **Blakemore, M.** (ed) *Proceedings of Auto Carto, London.* Royal Institute of Chartered Surveyors, pp. 572–7

Haines-Young, R. H. (1983) Nutrient cycling and problem solving: a simple teaching model. *Journal of Geography in Higher Education* 7: 125–39

Haines-Young, R. H., Petch, J. R. (1986) *Physical Geography: its Nature and Methods.* Harper and Row

Hall, D. K., Martinac, J. (1985) *Remote Sensing of Ice and Snow.* Chapman and Hall, New York

Hammond, C., Stobie, I. (1987) Desk-top publishing. *Practical Computing* January: 89–98

Hill, M. O. (1979a) *DECORANA – A FORTRAN program for Detrended*

Correspondence Analysis Reciprocal Averaging. Cornell University, Ithaca, New York

Hill, M. O. (1979b) *TWINSPAN – A FORTRAN program for arranging multi-variate data in an ordered two-way table by classification of the individuals and attributes.* Cornell University, Ithaca, New York

Hopgood, F. R. A., Duce, D. A., Gallop, J. R., Sutcliffe, D. C. (1983) *Introduction to the Graphics Kernel System (GKS).* Academic Press, New York

Horner, A. A., Walsh, J. A., Williams, J. A. (1984) *Agriculture in Ireland: a Census Atlas.* Dublin University

Hudson, R., Rhind, D. W., Mounsey, H. (1984) *An Atlas of EEC Affairs.* Methuen, New York

Huff, D. (1973) *How to Lie with Statistics.* Penguin

Imhoff, M. L., Petersen, G. W., Sykes, S. G., Irons, J. R. (1982) Digital overlay of cartographic information on Landsat MSS data for soil surveys. *Photogrammetric Engineering and Remote Sensing* 48: 1337–42

James, M. (1983) *Pascal for Micros.* Butterworth

Jensen, J. R. (1986) *Introductory Digital Image Processing: a Remote Sensing Perspective.* Prentice-Hall, Englewood Cliffs, New Jersey

Jensen, J. R., Hodgson, M. F., Christensen, E., Mackay, H. E., Tinney, L. R., Sharitz, R. (1986) Remote sensing inland wetlands: a multispectral approach. *Photogrammetric Engineering and Remote Sensing* 52: 87–100

Jensen, J. R., Ramsey, E. W., Mackay, H. E., Christensen, E. J., Sharitz, R. R. (1987) Inland wetland change detection using aircraft MSS data. *Photogrammetric Engineering and Remote Sensing* 53: 521–9

Johannsen, C. J., Sanders, J. L. (1982) (eds) *Remote Sensing for Resource Management.* Soil Conservation Society of America, Ankey, Iowa

Johnston, R. J. (1980) *Multivariate Statistical Analysis in Geography: a Primer on the General Linear Model.* Longman

Johnston, R. J. (1987) *Geography and Geographers: Anglo-American Human Geography since 1945* 2nd edn. Edward Arnold

Jones, A. R. (1985) The MICROMAP computer-assisted cartography package. *Computers & Geosciences* 11: 319–24

Jones, T. (1985) Mapping packages. In **England, J.** *et al.* (eds) *Information Systems for Policy Planning in Local Government.* Longman, pp. 378–95

Kememy, J., Kurtz, T. (1985) *Back to Basics: the History, Corruption and Future of the Language.* Addison-Wesley

Kent, A. (1983) (ed) *Geography Teaching and the Micro.* Longman

Kent, A. (1987) (ed) *Computers in Action in the Geography Classroom.*
Geographical Association

King, R. (1981) (ed) Theme issue on games and simulations in
geography teaching. *Journal of Geography in Higher Education* **5**: (2)

Kirkby, M. J., Burt, T. P., Naden, P. S., Butcher, D. P. (1987)
Computer Simulation in Physical Geography. Wiley

Landis, J. D. (1985) Electronic spreadsheets in planning. *Journal of
the American Planning Association* **51**: 216–24

Lang, K. (1985a) For the record. *Business Computing: the Survival
Game. Personal Computer World Special*, pp. 63–6

Lang, K. (1985b) ViewStore. *Personal Computer World* October: 194–
97

Lang, K. (1985c) Text play. *Business Computing: the Survival Game.
Personal Computer World Special*, pp. 50–5

Laurie, P. (1983) *Databases: How to Manage Information on your Micro.*
Chapman and Hall/Methuen

Lee, J. D., Lee, T. D. (1982) *Statistics and Computer Methods in
BASIC.* Van Nostrand Reinhold

Lee, J. D., Lee, T. D. (1986) Kermit. *Practical Computing* **9** (5): 114–
15

Lee, M. P., Soper, J. B. (1987) Using spreadsheets to teach statistics
in geography. *Journal of Geography in Higher Education* **11**: 27–
33

Letcher, P. (1985) *Which Peripherals? How to Choose Them, How to Use
Them.* Chapman and Hall/Methuen

Levine, N. (1985) The construction of a population analysis program
using a microcomputer spreadsheet. *Journal of the American Planning
Association* **51**: 496–511

Lewis, M. (1986) Practibase: trouble on the menu. *Practical Computing*
9 (3): 71

Lillisand, T. M., Kiefer, R. W. (1987) *Remote Sensing and Image
Interpretation* 2nd edn. Wiley, New York

Lo, C. P. (1976) *Geographical Applications of Aerial Photographs.* David
and Charles; Crane, Russak and Company Inc., New York

Lo, C. P. (1986) *Applied Remote Sensing.* Longman

McTaggart, L. (1985) WordStar 2000. *What Computer?* Spring: 25–6

MacDougall, E. B. (1983) *Microcomputers in Landscape Architecture.*
Elsevier, New York

Maguire, D. J. (1985a) Computers and cartography: space the final
frontier? *Bulletin of the Society of University Cartographers* **19**: 33–5

Maguire, D. J. (1985b) Microcomputer graphics. *Bulletin of the Society
of University Cartographers* **19**: 94–6

Maguire, D. J. (1986a) Computer-drawn statistical graphics. *Bulletin
of the Society of University Cartographers* **20**: 36–8

Maguire, D. J. (1986b) Generalization, fractals and spatial databases. *Bulletin of the Society of University Cartographers* **20**: 96–9

Maguire, D. J., Brayshay, W., Chalkley, B. S. (1987) *Plymouth in Maps: a Social and Economic Atlas.* Plymouth Polytechnic, 94

Maguire, D. J., Mingins, P. S., Saunders, M. N. K., Whitelegg, J. (1983) *Lancaster District: a Computer-drawn Census Atlas (1981).* University of Lancaster

Maguire, D. J., Mingins, P. S., Saunders, M. N. K., Whitelegg, J. (1984) Production of a census atlas by computer: Lancaster District (1981). *Bulletin of the Society of University Cartographers* **11**: 17–24

Maguire, D. J., Mohan, J. F. (1985) Towards a geographical information system for health care planning: harnessing a breakthrough to meet the needs of health care. *Health and Social Services Journal* 9 May: 580–1

Mandlebrot, B. B. (1977) *Fractals: Form, Chance and Dimension.* Freeman, San Francisco

Marble, D. F., Calkins, H. W., Peuquet, D. J. (1984) (eds) *Basic Readings in Geographic Information Systems.* SPAD Systems, Williamsville, New York

Marshall, G. J. (1983) *Programming Languages for Micros.* Butterworth

Martin, J. (1984) *Fourth Generation Languages. Volume II: Survey of Representative 4GLs.* Savant

Mather, P. M. (1976) *Computers in Geography: a Practical Approach.* Blackwell

Meadows, D. H., Meadows, D. L., Randers, J., Behrens III, W. (1973) *The Limits to Growth.* Universe Books, New York

Megarry, J. (1985) *Inside Information: Computers, Communications and People.* British Broadcasting Corporation

Meteorological Office (1982) *Observers Handbook.* HMSO

Midgley, H., Walker, D. (1985) *Microcomputers in Geography Teaching.* Hutchinson

Millington, A. C., Townshend, J. R. G. (1987) The potential of satellite remote sensing for geomorphological investigations – an overview. In **Gardiner, V.** (ed) *International Geomorphology 1986 II.* Wiley, pp. 331–42

Minnesota Department of Energy Planning and Development (1983) *Minnesota in the Eighties.* Division of Planning Department of Energy Planning and Development, St Pauls, Minnesota

Minnesota Land Planning Information Center (1987) *Introducing the Minnesota Land Management Information Center.* Planning Information Center, St Pauls, Minnesota

Minshull, R. (1975) *An Introduction to Models in Geography.* Longman

Moffatt, I. (1986) Teaching environmental systems modelling using computer simulation. *Journal of Geography in Higher Education* **10**: 53–60

Moller, A. (1984) *LOGO Programming: a Practical Guide for Parents and Teachers.* Century Communications

Monmonier, M. S. (1982) *Computer-aided Cartography: Principles and Prospects.* Prentice-Hall, Englewood Cliffs, New Jersey

Moody, G. (1985) Getting down to business. *Practical Computing* **8** (9): 85–7

Morgan, R., Wood, B. (1982) *Word Processing.* Longman

Morrison, J. L. (1986) Cartography: a milestone and its future. In **Blakemore, M.** (ed) *Proceedings of Auto Carto, London Vol 1.* Royal Institute of Chartered Surveyors, pp. 1–12

Moser, C. A., Scott, W. (1961) *British towns: a statistical study of their social and economic differences.* Oliver and Boyd

Murray, M. A. (1974) *Atlas of Atlanta: the 1970s.* The University of Alabama Press, Alabama

Neelamkavil, F. (1987) *Computer Simulation and Modelling.* Wiley

Negus, J. (1985) FACTORY LOCATION – a review. *Teaching Geography* **10**: 188

Newman, W. M., Sproul, R. F. (1979) *Principles of Interactive Computer Graphics.* McGraw-Hill, Japan

Nichol, J., Briggs, J., Dean, J. (1987) Authoring programs and toolkits, logic programming and curriculum development. In **Kent, W. A., Lewis, R.** (eds) *Computer-assisted Learning in the Humanities and Social Sciences.* Blackwell, pp. 151–63

North, P. F. (1983) A computer-based system for the acquisition and analysis of data from a field tillage study. *Computers & Geosciences* **9**: 229–34

O'Brien, L. G. (1986) Statistical software for microcomputers. *Area* **18**: 39–42

O'Brien, L. G., Wrigley, N. (1980) Computer programs for the analysis of categorical data. *Area* **12**: 263–8

O'Keefe, L., Klagge, J. (1986) *Statistical Packages for the IBM PC Family.* McGraw-Hill, New York

Openshaw, S. (1987) Research policy and review 17. Some applications of supercomputers in urban and regional analysis and modelling. *Environment and Planning A* **19**: 853–60

Openshaw, S., Rhind, D. W., Goddard, J. (1986) Geography, geographers and the BBC Domesday Project. *Area* **18**: 9–13

Ottensmann, J. R. (1985) *BASIC Microcomputer Programs for Urban Analysis and Planning.* Chapman and Hall, New York

Palmer, J. J. N. (1986) Computerizing Domesday Book. *Transactions of the Institute of British Geographers* **11**: 279–89

Pearson, A., Sprunt, B. (1987) A microcomputer-based map retrieval system. *Bulletin of the Society of University Cartographers* **20**: 100–102

Pease, R. W. (1987) The average surface temperature of the earth: an energy budget approach. *Annals of the Association of American Geographers* **77**: 450–61

Perry, J. T., McJunkins, R. F. (1985) *A User's Guide to dBASE II.* Blackwell

Persand, K. C., Virden, R. (1984) Data-logging with microcomputers. In **Ireland, C. R., Long, S. P.** (eds) *Microcomputers in Biology: a Practical Approach.* IRL Press, pp. 43–65

Peucker, T. K., Chrisman, N. (1975) Cartographic data structures. *The American Cartographer* **2**: 55–69

Reeve, D. E. (1985) Computer mapping: mainframe to micro, research to classroom. *Computers & Geosciences* **11**: 313–18

Reeve, D. E., Carrick, R. J. (1983) Social atlas of Kirklees. *Local Information Paper 6.* Department of Geography, Huddersfield Polytechnic

Reif, B. (1973) *Models in Geography and Regional Planning.* Leonard Hill

Rhind, D. W. (1977) Computer-assisted cartography. *Transactions of the Institute of British Geographers* **2**: 71–97

Rhind, D. W. (1983) Towards a national digital topographic data base: experiments in mass digitizing, parallel processing and the detection of change. In **Weller, B. S.** (ed) Automated cartography: international achievements and challenges. *Proceedings of the 6th International Symposium on Automated Cartography*, pp. 428–37

Rhind, D. W. (1983) (ed) *A Census User's Handbook.* Methuen

Rhind, D. W. (1985) Geographical data handling: recent developments. *Computers & Geosciences* **11**: 297–8

Rhind, D. W. (1988) Computing, academic geography and the outside world. In **MacMillan, W.** (ed) *Remodelling Geography.* Blackwell, Oxford

Rhind, D. W., Wyatt, B., Briggs, D., Wiggins, J. (1986) The creation of an environmental information system for the European Community. *Nachrichten aus dem Karten und Vermessungswesen Series II.* Translations–44: 147–57

Roberts, N., Andersen, D., Deal, R., Garet, M., Shaffer, W. (1983) *Introduction to Computer Simulation: a Dynamics Modelling Approach.* Addison-Wesley, Reading, Massachusetts

Robinson, A. H., Sale, R. D., Morrison, J. L., Muehrcke, P. C. (1984) *Elements of Cartography* 5th edn. Wiley, New York

Robinson, V. B., Frank, A. U., Matthew, A. B. (1986) Expert systems applied to problems in geographic information systems:

introduction, review and prospects. *Computers Environment and Urban Systems* 11: 161–73

Rosen, A. (1983) *Getting the Most out of your Word Processor.* Prentice-Hall, Englewood Cliffs, New Jersey

Rowe, B. C., Cable, D., Neffendorf, H., Westlake, A. J. (1985) Statistical analysis packages. In **England, J. R., Hudson, K. I., Master, R. J., Powell, K. S., Shortridge, J. D.** (eds) *Information Systems for Policy Planning in Local Government.* Longman, pp. 364–77

Rowley, G. (1985) Developments in survey research. *Area* 17: 115–16

Rowley, G., Barker, K., Callaghan, V. (1985) The Questronic Project and the Ferranti MRT 100: a boon for survey research. *The Professional Geographer* 37: 459–63

Rudeforth, C. C. (1982) Handling soil survey data. In **Bridges, E. M., Davidson, A. D.** (eds) *Principles and Applications of Soil Geography.* Longman, pp. 97–131

Ryan, B. F., Joiner, B. L., Ryan, T. A. (1985) *Minitab Handbook* 2nd edn. Duxbury Press, Boston

Sabbins, F. F. (1987) *Remote Sensing: Principles and Interpretation* 2nd edn. Freeman, New York

Saunders, R. W., Ward, N. R., England, C. F., Hunt, G. E. (1982) Satellite observations of sea temperature around the British Isles. *Bulletin of the American Meteorological Society* 63: 267–72

Sawicki, D. S. (1985) Microcomputer applications in planning. *Journal of the American Planning Association* 51: 209–15

Seyer, M. D. (1984) *RS232 Made Easy: Connecting Computers, Printers, Terminals and Modems.* Prentice-Hall, Englewood Cliffs, New Jersey

Shain, M. (1986) Learning to lock up your data. *PC* 3: 56–75

Sharitz, R. (1986) Remote sensing inland wetlands: a multispectral approach. *Photogrammetric Engineering and Remote Sensing* 52: 87–100

Sharp, J. J., Sawden, P. (1984) *BASIC Hydrology.* Butterworth

Shaw, G., Wheeler, D. (1985) *Statistical Techniques in Geographical Analysis.* Wiley

Sheath, N. J. (1986) Development of an integrated map plotting system for hydrocarbon exploitation. In **Blakemore, M.** (ed) *Proceedings of Auto Carto, London Vol 1.* Royal Institute of Chartered Surveyors, pp. 66–75

Shelley, J., Hunt, R. (1984) *Computer Studies: a First Course* 2nd edn. Pitman

Shepherd, I. D. H. (1985) Teaching geography with the computer: possibilities and problems. *Journal of Geography in Higher Education* 9: 3–23

Shepherd, I. D. H., Cooper, Z. A., Walker, D. R. F. (1980)

Computer-assisted Learning in Geography: Current Trends and Future Prospects. Council for Educational Technology with the Geographical Association

Shih, E. H. H., Schwengerdt, R. A. (1983) Classification of arid geomorphic surfaces using spectral and textural features. *Photogrammetric Engineering and Remote Sensing* **49**: 337–47

Short, N. M. (1982) *The Landsat Tutorial Workbook: Basics of Satellite remote sensing.* NASA, Washington DC

Silk, J. (1979) *Statistical Concepts in Geography.* Allen and Unwin

Sipe, N. G., Hopkins, R. W. (1984) *Microcomputers and Economic Analysis: Spreadsheet Templates for Local Government.* Bureau of Economic and Business Research, University of Florida, Gainsville

Smith, T. R. (1984) Artificial intelligence and its applicability to geographical problem solving. *The Professional Geographer* **36**: 147–58

Smith, T. R., Menon, S., Star, J. L., Estes, J. E. (1987) Requirements and principles for the implementation and construction of large-scale geographic information systems. *International Journal of Geographical Information Systems* **1**: 13–31

Someren, A. V. (1986) With a view to storage. *Acorn User* May: 127–9

Soper, J. B., Lee, M. P. (1987) *Statistics with Lotus 1–2–3.* Chartwell Bratt

Southard, R. B. (1987) Automation in cartography – revolution or evolution. *The Cartographic Journal* **24**: 59–63

Sparks, L. (1983) *Final report of the Microcomputer Environmental Probes Project.* Saint David's University College

Sparks, L., Sumner, G. (1982) Monitoring weather by computer. *Educational Computing* **4**: 19–20

Sparks, L., Sumner, G. (1984a) Micros in control – on-line weather data acquisition using a BBC microcomputer. *Weather* **39**: 212–18

Sparks, L., Sumner, G. (1984b) A microcomputer-based weather station monitoring system. *Journal of Microcomputer Applications* **7**: 233–42

Stamp, L. D. (1938) *A Regional Geography for Higher Certificate and Intermediate Courses: Part V Europe and the Mediterranean* 7th edn. Longman

Starr, L. E. (1986) Mark II: the next step in digital systems development at the US Geological Survey. *The American Cartographer* **13**: 368–71

Stephenson, A. P., Stephenson, D. J. (1984) *Filing Systems and Databases for the BBC Micro.* Granada

Sterling, J. A. L. (1986) *The Data Protection Act 1984: a Guide to the New Legislation* 2nd edn. Tax and Business Law

Stewart, A. (1987) Handling spatial data – GEOBASE. In Kent, A. (ed) *Computers in the Classroom.* Geographical Association, pp. 8–10

Stobie, I. (1985) Filevision. *Practical Computing* 8 (1): 104–5

Stoffel, D. B., Stoffel, K. L. (1980) Mt St Helens seen close up on May 18. *Geotimes* 25: 16–17

Suguira, R., Sabins, F. (1980) The evaluation of 3 cm wavelength radar for mapping surface deposits in the Bristol Lake/Granite Mountain area, Mojave Desert. *Radar Geology: an Assessment*. JPL, Passedena, California, 439–56

Sumner, G., Sparks, L. (1984) Lightning never strikes twice! *Area* 16: 109–14

Taylor, D. R. F. (1984) Computer-assisted cartography, new communication technologies and cartographic design: the need for a new cartography. *Technical Papers of the 12th Conference of the International Cartographic Association, Perth* 2: 457–66

Tennant-Smith, J. (1985) *Basic Statistics*. Butterworth

Teskey, F. N. (1982) *Principles of Text Processing. Computers and their applications 15*. Ellis Horwood

Thomas, R. W., Huggett, R. J. (1980) *Modelling in Geography: a Mathematical Approach*. Harper and Row

Tobler, W. R. (1959) Automation and cartography. *Geographical Review* 49: 526–34

Tomlinson, R. F. (1984) Geographic Information Systems – a new frontier. *Proceedings of the International Symposium on Spatial Data Handling*, Zurich, Switzerland, 1–14

Townsend, A., Blakemore, M., Nelson, R., Dodds, P. (1986) The National On-line Manpower Information System (NOMIS). *Employment Gazette* 94: 60–4

Townsend, A., Blakemore, M., Nelson, R. (1987) The NOMIS data base: availability and uses for geographers. *Area* 19: 43–50

Tukey, J. W. (1977) *Exploratory Data Analysis*. Addison-Wesley, Reading, Massachusetts

Unwin, D. J. (1974) Hardware provision for quantitative geography in the United Kingdom. *Area* 4: 200–4

Unwin, D. J. (1981) *Introductory Spatial Analysis*. Methuen

Unwin, D. J., Dawson, J. A. (1985) *Computer Programming for Geographers*. Longman

Vincent, P., Haworth, J. (1984) Statistical Inference: the use of the likelihood function. *Area* 16: 153–8

Walford, R. (1981) Geography games and simulations: learning through experience. *Journal of Geography in Higher Education* 5: 113–19

Walsh, V. (1985) *Computer Literacy: a Beginner's Guide*. Macmillan

Watson, D. (1984a) Microcomputers in secondary education – a perspective with particular reference to the humanities. In **Kelly, A. V.** (ed) *Microcomputers and the Curriculum*. Harper and Row, pp. 125–44

Watson, D. (1984b) (ed) *Exploring Geography with Microcomputers. MEP Readers 3*. Council for Education Technology

Waugh, T. C., McCalden, J. (1983) *GIMMS Reference Manual: Release 4.5*. GIMMS Ltd

Weatherill, G. B., Curram, J. B. (1984) *The Design and Analysis of Statistical Software for Microcomputers*. University of Kent

Webb, R. P. (1982) A synopsis of natural resource management and environmental assessment techniques using geographical information systems technology. *Computers Environment and Urban Systems* 7: 219–31

Wells, M. (1986) A progress report on JANET. *University Computing* 8: 146–53

Wiggins, J. C., Hartley, R. P., Higgins, M. J., Whittaker, R. J. (1987) Computing aspects of a large geographic information system for the European Community. *International Journal of Geographical Information Systems* 1: 77–87

Wiggins, L. L. (1986) Three low-cost mapping packages for microcomputers. *Journal of the American Planning Association* 52: 480–88

Wilding, C. M. (1985) From cards to computers: the SNAP market survey analysis package. *16 bit Computing* May: 1–5

Williams, R. G. B. (1985) *Intermediate Statistics for Geographers and Earth Scientists*. Macmillan

Wrigley, N. (1985) *Categorical Data Analysis for Geographers and Environmental Scientists*. Longman

Zorkoczy, P. (1985) *Information Technology: an Introduction* 2nd edn. Pitman

Glossary

Algorithm	A clearly defined sequence of steps which, if followed, will provide a guaranteed solution to a specific problem.
Analogue	A continuous form which in theory can take any value throughout its range, e.g. voltage or angular position.
Analogue to digital converter	A device which converts analogue signals into a digital form. Also called an A/D converter.
ASCII	Acronym for American Standard Code for Information Interchange. An 8 bit coding system used to represent alphanumerical characters in computers.
Assembler	A program which converts assembly language programs into machine code.
Assembly language	A low level programming language which uses mnemonics to represent instructions.
Bar code reader	A device for reading bar codes – binary combinations of thick and thin vertical black bars used to mark foods, library books etc.
Baud	A unit used to measure the transmission speed of digital data which is approximately the same as bits per second.
Binary	A number system which uses the base 2 and has only the digits 0 and 1.

Bit

BInary digiT. A bit can have only two states, 0 and 1.

Byte (b)

The unit which describes the number of bits used to handle a single character. Some people prefer to use the term word for this purpose and restrict byte to a group of 8 bits.

Cathode Ray Tube (CRT)

An electronic tube (similar to those used in televisions) in which a beam of electrons can be directed onto a phosphor-coated surface to produce visible information. The principal component of the majority of computer screens.

Central Processor Unit (CPU)

The main part of any computer, which fetches and carries out program instructions; comprises an arithmetic logic unit (ALU), a control unit (CU) and a memory unit (MU).

Chip

The common name for an integrated circuit.

Compiler

A program which converts a high level language program (source code) into machine code (object code) and checks the overall logic of the program. See also Interpreter.

Computer

A programmable electronic machine which can input, store, manipulate and output data.

Coprocessor

Any additional device which carries out some or all of the operations of a standard processor. Also called a second processor.

Cursor

A marker on a screen which appears at the position where an operation will take place. Usually a flashing square or underscore.

Data

A general term that can describe numbers, characters or groups of bits suitable for processing by a computer.

Data base

A large integrated collection of data.

Data Base Management System (DBMS)

A program for creating, maintaining and managing a data base.

Data logger

An automatic or semiautomatic device used to collect and record data.

Digital

A discrete form which can only take specific individual values throughout its range. Digital computers are two state devices in which data can be represented by combinations of 0 and 1.

Digital to analogue converter	A device which converts digital signals into an analogue form. Also called a D/A converter.
Disk drive	The basic storage device of a computer system. The hardware device which transfers data between a disk and main memory.
Disk Operating System (DOS)	The software which controls the transfer of data between disks and main memory.
Erasable Programmable Read Only Memory (EPROM)	A ROM memory chip which can be reprogrammed several times after erasing the contents using ultra-violet light.
Field	A subdivision of a record which contains one unit of information, for example, the answer to a single question from a questionnaire.
File	A collection of related data stored on a computer.
Flat file	A file in which the records have the same structure (same number of fields etc.) and are not related to the records in another file.
Floppy disk	A circular flexible plastic-coated metal oxide disk enclosed in a plastic sleeve on which data can be stored.
Hand-held portable terminal	A type of portable, usually battery-powered device, into which data can be keyed and later transferred to a computer.
Handshaking	A widely used communication protocol for controlling data transmission.
Hard disk	See Winchester disk.
Hardware	The physical equipment in a computer system.
Hexadecimal	A number system which uses the base 16. It is particularly used to represent long binary numbers in a shorter and more understandable form.
Input/output interface	The part of a microprocessor which acts as a link between the central processor unit and the peripherals of a microcomputer.
Integrated circuit	A circuit package that contains several components chemically formed upon a single piece of semiconductor material; commonly called a chip.
Interpreter	A program which converts a high level language

program line-by-line into machine code and immediately executes the instructions.

Keyboard	The basic input device of a computer system which is rather like a typewriter.
Kilobyte (kb)	1 kb = 1,024 bytes.
Language	A defined set of commands which can be used to program a computer, examples include BASIC, Pascal, FORTRAN and BCPL.
Liveware	The people necessary to operate and use computers.
Machine code	The coding system used to represent instructions which can be directly recognized by a computer.
Macro	A set of instructions (commands) which can be executed collectively by giving the name of the macro.
Megabyte (mb)	1 mb = 1,024 kb = 1,048,576 bytes.
Microcomputer	A type of computer in which the functions of the central processor unit are carried out by one or a few microprocessors.
Microfilm/ microfiche plotter	An output device which plots high quality images on film. Particularly suitable for information which requires only occasional reference.
Microprocessor	A Central Processor Unit (CPU) constructed on a single Very Large Scale Integration (VLSI) integrated circuit.
Modem	A contraction of the term MOdulator–DEModulator. A type of analogue to digital/digital to analogue converter which allows computers to communicate via the telephone network.
Mouse	A device used to move a screen cursor and input commands; commonly used in graphics.
Nibble	1 nibble = 4 bits.
Operating System (OS)	The software which controls all the operations of a microcomputer.
Optical character reader	A device capable of reading text directly into a computer. Most are expensive and require special typefaces, but they are considerably quicker than most other data entry methods.

Package	A generalized program capable of performing several operations and covering the requirements of many users.
Paddles	Rotating knobs, which control the movement of a cursor in either an *x* or *y* direction. Normally used in pairs; less common than joysticks, mice and trackballs.
Peripheral	Any device added to a computer system.
Pixel	An individual dot on a screen. Also known as a pel which is derived from Picture ELement.
Program	A set of processing steps that a computer is required to perform.
Programmable Read Only Memory (PROM)	A ROM memory chip which can be programmed once using a special PROM programmer device.
Protocol	A technical convention; widely used to establish communication links.
Random Access Memory (RAM)	A type of chip-based memory which a processor (CPU) can 'read' data from and 'write' data to.
Read Only Memory (ROM)	A type of chip-based memory which a processor (CPU) can 'read' data from but cannot 'write' data to.
Record	A collection of related fields, for example, all the responses from one questionnaire.
Resolution	The resolution of a computer is normally expressed as the number of pixels which can be displayed along the *x* and *y* axes of a display.
Robot arm	A mechanical device controlled by a computer which is capable of doing the work of a human arm.
RS232–C	The most common form of communication interface protocol used in computer systems. The RS422–A and RS423–C are derivatives of it.
Screen	The basic output device of a computer system. The majority of screens are based on Cathode Ray Tubes (CRT) although some recent screens have Liquid Crystal Displays (LCDs) and Electroluminescent Displays (EDs). Also called a monitor, and Visual Display Unit (VDU).
Speech synthesizer	Any device that produces sound patterns which simulate speech.
Spreadsheet	A type of software which allows repeated

calculations to be performed on sets of data. In a spreadsheet rows and columns of numbers are linked together by formulae. A change in any cell in a spreadsheet automatically results in the updating of row and column totals and other related statistics.

Software   Computer programs and data files.

Teletext   A term used to describe non-interactive, that is read only, videotex systems.

Touch Screen   A device where users point at areas on a screen and infrared beams detect the relevant spot and act accordingly.

Trackball   A rotating ball in a socket which when turned activates potentiometers that produce digital co-ordinates. Used to control cursor speed and direction.

VDU   See Screen.

Very Large Scale Integration (VLSI)   The process of locating over 10,000 transistors on a single chip.

Videotex   A term used to describe computer-based information storage and retrieval systems which display information in the form of pages on a screen. Viewdata and teletext are the two forms of videotex.

Viewdata   A term used to describe interactive videotex systems.

Voice recognition system   Any device which can recognize spoken words. At present these are not widely used because they are either expensive or have very limited vocabularies.

Winchester disk   A non-removable disk on which data can be stored. Although similar in principle to floppy disks they have a higher capacity, faster access time and are more reliable.

Window Icon Mouse Pull-down menu Systems (WIMPS)   A type of extension to a microcomputer operating system which acts as a graphical interface between the user and the computer.

Word   The basic unit for handling data in a computer system.

Index

Items in **bold** indicate major entry.

Computers in Geography reflects the enormous contribution that computers have made to geographical study during the 1980s. It considers how geographers from all branches of the discipline can enhance their work by using computers.

It is not a theoretical computer manual, indeed it assumes little knowledge of computers and quantitative techniques. Examples are used throughout to demonstrate the application of information technology to specific problems in both human and physical geography. There is discussion of how computers can contribute to all stages of the process of geographical explanation and this covers consideration of data collection, storage, management, analysis and presentation. The book includes a discussion of the exciting new developments in geographical information systems. The text is generously illustrated throughout with over 100 photographs, drawings and tables and a glossary is provided to explain the technical phrases used within.

Primarily intended for undergraduate courses in the principles and applications of computers in geography, this text will also provide supplementary reading for students of biological, environmental and social sciences.

David J. Maguire is Lecturer in Geography at the University of Leicester.

Cover illustration: Andrew Lovell

ISBN 0-582-30171-8

9 780582 301719

Longman
Scientific &
Technical

Copublished in the United States with
John Wiley & Sons, Inc., New York
ISBN 0–470–21194–6 (USA only)